OXFORD STUDIES IN AFRICAN AFFAIRS

General Editors
JOHN D. HARGREAVES *and* GEORGE SHEPPERSON

RAILWAYS AND THE COPPER MINES OF KATANGA

SIR ROBERT WILLIAMS, BART.

The portrait hanging in the Boardroom of the London Offices of Tanganyika Concessions, Ltd.

RAILWAYS AND THE COPPER MINES OF KATANGA

BY

S. E. KATZENELLENBOGEN

CLARENDON PRESS · OXFORD
1973

Oxford University Press, Ely House, London W.1

GLASGOW NEW YORK TORONTO MELBOURNE WELLINGTON
CAPE TOWN IBADAN NAIROBI DAR ES SALAAM LUSAKA ADDIS ABABA
DELHI BOMBAY CALCUTTA MADRAS KARACHI LAHORE DACCA
KUALA LUMPUR SINGAPORE HONG KONG TOKYO

ISBN 0 19 821676 9

PRINTED IN GREAT BRITAIN
BY RICHARD CLAY (THE CHAUCER PRESS) LTD
BUNGAY, SUFFOLK

TO MY MOTHER AND FATHER

Preface

ONE of the obstacles to an understanding of economic development in Africa during the colonial period is the impossibility of finding source material covering all the political, financial, and personal aspects of it. The problem is not unique to the economic historian, but is exacerbated by the difficulty of locating and gaining access to business records and the private papers of the individuals concerned. Companies operating in Africa have in recent years come under heavy criticism, and it is not really surprising that businessmen and their heirs are somewhat reticent about allowing someone to investigate their pasts. In addition to varying attitudes in the business world towards the preservation of archives and allowing access to them, there are different government policies controlling access to official records. Britain, for example, opens most department files after thirty years, although the long-term advantage of this liberal policy for historians is very much open to question. Belgium and France operate a fifty-year rule, while Portugal considers individual requests for permission to consult archives, but has not as yet allowed anyone to see papers relating to Africa in the twentieth century. It is this imbalance of material that has forced me to treat the construction of railways within the Congo itself much more briefly than the Benguela and Rhodesian railways.

As a general rule, people in the world of finance seem to be much more discreet than their counterparts in politics or the diplomatic and civil services. Memoirs by businessmen are rare, so it was with great anticipation that I began reading those of Albert Pam, a London banker involved in one of the unsuccessful attempts to finance the construction of the Benguela Railway during the inter-war years, only to find in an introductory note: 'I have written nothing about my business life, although I have been asked to do so. But life in the City of London is too intimate and personal, and I would have to mention the names of firms and individuals at every turn. It is better to avoid this.'[1] Pam goes on to describe hunting trips and travels around the world, making only one passing reference to a business colleague.

[1] Albert Pam, *Adventures and Recollections* (Oxford University Press, privately for Pam, Oxford, 1945), [p. iii].

The Countess of Warwick, more involved in financial affairs than she cared to admit, could have shed considerable light on developments in central Africa, but ends one volume of her memoirs saying: 'Some friends seemed to think I had much to tell, I have not, for I have shared too many confidences, have known so much that has gone on behind the scenes. Were I once to let go where should I stop, how many friends betray? So I have avoided personalities.'[2] This reticence is understandable, but a hindrance none the less.

One point which must be accepted if the activities of businessmen in Africa are to be analysed with any degree of objectivity is that these were 'men whose standards were those of the mining camp, the stock exchange and the gaming table'.[3] Relations between them were often rough, and personal or business loyalties, family connections, or similar sentiments could be invoked as justification for actions, but could not be allowed to interfere with the fundamental aim of avoiding financial loss where possible and making as great a profit as they could. It is against this ethic that I have tried to determine the success or failure of the various enterprises connected with the exploitation of Katanga's copper resources. The question of whether or not the actions of these men and their companies were good in any wider moral sense or in regard to their impact on Africa and Africans is far from irrelevant, but is beyond the scope of this book.

In the various stages of this work I have enjoyed the help and advice of many people. Professor Jean Stengers of the Free University of Brussels first suggested the topic to me and gave me the benefit of his great knowledge of the Congo's history, of the personalities involved in it, and of archives in both Belgium and Britain. I am very much indebted to the directors, officers, and staff of Tanganyika Holdings Limited who gave me free access to material in their possession and the hospitality of their offices during the many months it took me to go through their files. My thanks go particularly to Messrs. R. J. MacNeile, A. C. Herrick, J. Ralph and Norman Munn; also to Mr. S. G. Watts, the Benguela Railway Company's London Manager, and to Shr. Louis Lane, its London Secretary. Others in Britain helped me in connection with source material: Mr. Airey Neave lent me his grandfather's diaries; Mr. I. J. C. Foster and the

[2] Frances, Countess of Warwick, *Life's Ebb and Flow* (Hutchinson, London, 1929), p. 273.

[3] J. van der Poel, *The Jameson Raid* (Oxford University Press, Cape Town, 1951), p. 3.

Cabinet Office allowed me to consult and cite the Wingate Papers; the Librarian of Rhodes House, Oxford, permitted me to consult the Rhodes and Cawston Papers; the late Col. Phillip Gell and Mrs. Aileen Gell, O.M., gave me access to the papers of Philip Lyttelton Gell, and Gwladys, Lady Norton-Griffiths, let me read the autobiography she wrote for circulation among her friends and family, and shared some of her reminiscences of the time she spent in Angola with her husband during the early stages of the Benguela Railway's construction; Mr. Frank Tiarks provided me with material from the files of Schröder, Wagg & Company.

In Belgium I benefited greatly from the help and advice of many people: Mme de Smet of the Archives Générales du Royaume, M. Emile Van de Woude, Archivist of the Belgian Royal Palaces, Mme Madeleine van Grieken-Taverniers, archivist of the Foreign Ministry's African section (formerly the Colonial Ministry); M. Ph. Desneux, Foreign Ministry archivist; MM. M. Walraet and A. Bourgeois, formerly of the Comité Spécial du Katanga; M. Lucien Cahen, Director of the Musée Royale de l'Afrique Centrale at Terveuren, and the curator of the museum's historical section, M. M. Luwel who in addition to wide knowledge and charm has the distinction of being able to read the handwriting of King Leopold II with apparent ease.

Others to whom I should like to express my thanks for their assistance are Professor J. A. Gallagher, late of Oxford University, who supervised several stages of this work in its original form as a doctoral thesis, Professor Roger Anstey, University of Kent, Canterbury, Professor S. H. Frankel, Nuffield College, Oxford, Professor Roland Oliver, School of Oriental and African Studies, Professor John Flint, formerly of King's College, London; Dr. Shula Marks, Institute of Commonwealth Studies, University of London, Mr. David Fieldhouse, Nuffield College, Oxford, M. Ph. Chevalier of the Société d'Investissement des Fonds du BéCéka (Sibeka) and M. Paret of Union Minière du Haut-Katanga; also Mrs. Hester Marsden-Smedley, Mr. H. J. M. Hambrook, a former teaching colleague in the Congo, Mr. Keith Kyle, formerly of the BBC, and Mr. J. A. T. Lewis of the Zebra Trust. A heavy debt of gratitude is also owed the staff of the Public Record Office, particularly Mr. Norman Evans. The map was originally drawn by Mr. David Orme, cartographer at Leicester University.

I should also like to pay special tribute to the late Mr. George

Bennett who supervised most of my work in its original form. His encouragement and careful criticism were invaluable. My wife put up with much during the preparation of this book and made many helpful suggestions. The comments made by Professor J. D. Hargreaves, editor of this series, and my colleague at the University of Manchester, Dr. W. H. Chaloner, are also very much appreciated. Miss Pauline Voce, with the help of Miss Frances Street, undertook cheerfully the unenviable task of typing the final draft.

For financial support of my work my thanks go to the Trustees of the Beit Fund and of the George Webb Medley Fund at Oxford University, the Master of St. Catherine's College, Oxford, and the Institute of Commonwealth Studies of the University of London where I spent the year 1967/8 as a Dame Lillian Penson Junior Research Fellow.

Without the help of these people I could not have completed this work; responsibility for it is, however, entirely my own.

Marple, Cheshire, S. E. KATZENELLENBOGEN
November 1972

Contents

Frontispiece: Sir Robert Williams, Bart. The portrait hanging in the Boardroom of the London Offices of Tanganyika Concessions, Ltd. (Photograph by Norman Munn, courtesy of Tanganyika Holdings Ltd.)

Abbreviations

BCK	Compagnie de Chemin du fer du Bas-Congo au Katanga (or the railway line itself).
BR	File of the London Committee of the Benguela Railway Company.
BRC	Benguela Railway Company.
BSAC	British South Africa Company.
BT	Board of Trade.
CAB	British Cabinet Records.
CCCI	Compagnie du Congo pour le Commerce et l'Industrie.
CFK	Compagnie du Chemin de fer du Katanga (or the railway line itself).
CO	British Colonial Office (or Colonial Office Records).
CSK	Comité Spécial du Katanga.
DOT	Department of Overseas Trade.
EIC	État Indépendant du Congo (or Archives of the State).
FO	Foreign Office (or Foreign Office Records).
IO	India Office (or India Office Records).
IRCB	Institut Royal Colonial Belge.
MAE	Belgian Ministry of Foreign Affairs (or Ministry Archives).
RRL	Rhodesia Railways Limited.
SA	Sudan Archive.
T	Treasury Records.
TC	File of Tanganyika Concessions Ltd.
TCL	Tanganyika Concessions Ltd.
TFAAC	Trade Facilities Act Advisory Committee.
TFN	Tanganyika Concessions Ltd. file on negotiations with the Trade Facilities Act Advisory Committee.
UM	Tanganyika Concessions Ltd. files pertaining directly to activities of the Union Minière du Haut-Katanga.
UMHK	Union Minière du Haut-Katanga.
VEP	The Papers of Edmund van Eetvelde.
ZE	File of the Zambesia Exploring Company.
ZEC	Zambesia Exploring Company.

NOTE

For the sake of uniformity I have adopted throughout the more modern spelling of Benguela, as opposed to Benguella as it was spelt during much of the period covered by this work.

CHAPTER I

The Scramble for Katanga

HOWEVER great or small a role private business interests had played in initiating Europe's expansion into Africa in the late nineteenth century, they were of the utmost importance once the powers had staked their claims to territories and had to solve the practical problems of establishing and maintaining control over them. One point of colonial policy on which all the powers concerned agreed was that their new possessions were to be financially independent, meeting the costs of their administration and defence from their own resources, and in some instances, producing surpluses which could be used in the mother country. The expansion of Africa's agricultural and mineral production to meet the demand from the growing industries of Europe and America was the key to generating sufficient local revenue to make this possible. Economic development meant imports of manufactured goods on which duty was paid, employment opportunities for Africans who could then be taxed, licence fees for prospecting, land taxes, etc. Parsimonious treasuries were not prepared to provide the capital needed for this development, but looked instead to private investors willing to take the very high risk of losing their money in the little-known African interior in the hope of gaining correspondingly high profits.

In carrying on their activities, financiers and the companies they formed looked to government for assistance in a variety of ways ranging from police protection of plant and personnel and help in labour recruiting to diplomatic representation in dealings with foreign countries and support for financial schemes. Government's dependence on companies made some degree of co-operation inevitable, but there was considerable scope for conflict. Because government had to take account of over-all national and colonial policy, domestic political questions, humanitarian pressures, and the state of international relations in Europe, their interests did not always coincide with those of particular companies. An understanding of the course of Africa's economic development during the colonial period requires an analysis not only of the activities of companies—how they secured their funds, how they determined financial and commercial policy, etc.—but also their relations with government, official attitudes towards development in general, and how particular decisions to support, oppose, or remain indifferent to a company were made.

It is also essential to consider the personalities involved in development, their strengths and weaknesses, their abilities and ambitions, and their relationships with others. The history of South Africa and the Rhodesias, for example, would certainly have been very different if men like Cecil Rhodes and Sir Starr Jameson had not been involved in it, or had been different kinds of men. King Leopold II of the Belgians exerted an inordinate influence not only on the Congo but also on the entire African continent. At all levels individuals influenced events; their role was crucial, and must be understood.

In central and southern Africa, where the primary attraction for investment was the mineral resources known or believed to exist in the interior, relationships between companies and governments became very complex. Large amounts of money were needed to search for minerals, bring mines to production, and build railways to connect them to the coast. In the Portuguese colonies, German South West Africa, and the Congo it was necessary to rely on foreign —particularly British and French—capital, often giving rise to the fear that financial involvement might be used as an instrument of politics. Within companies, individuals of different nationalities could work harmoniously together, but could often try to improve their own positions at one another's expense. The extent to which governments identified themselves with these international financial conflicts varied widely according to circumstances.

The copper mines of Katanga are an interesting case in point, the variety of interests involved with their development in one way or another having probably been greater than in any other part of the continent. Geographically and culturally Katanga was more closely linked to neighbouring Rhodesia than to the rest of the Congo. The original impetus and much of the finance for its development came from British interests already operating in Rhodesia and South Africa. The first, and until 1928 the only, unbroken railway route between the mines and the coast passed through Rhodesia and Moçambique. Portuguese interests also became involved as the Benguela Railway was built across Angola in the face of German opposition. The Independent Congo State and its Belgian successors strove to keep Katanga out of the British orbit and to retain control of the $\frac{1}{2}$ per cent of the Congo's area which, by the time the country became independent in 1960, produced 45 per cent of the total value of its exports.

LEOPOLD STAKES HIS CLAIM

The origins of Katanga's history as a part of the Congo State can be found primarily in the personality of King Leopold; an ambitious man, dissatisfied with being the monarch of a small uninfluential country relying for its continued existence on the goodwill of neighbouring great powers. The King wanted to make Belgium sufficiently strong to protect its own neutrality, and a political and economic force in the world. He was passionately interested in economic questions—often to the boredom of the civil servants and others whom he interrogated at length—and worked hard to encourage the expansion of Belgium's foreign trade. Not long before ascending the throne in 1865, he presented the Finance Minister, Walthère Frère-Orban, with a bas-relief portrait of himself carved in a piece of marble from the Acropolis, his head surrounded by the words 'Il faut à la Belgique une colonie'. The minister and most other Belgians profoundly disagreed with their monarch on this point, but he set out to acquire a colony for them whether they wanted it or not. He conceived a scheme in 1869 for acquiring a concession from Spain in the Philippines but by 1875 decided that Spain, Portugal, and the Netherlands had no overseas territory to sell him, and turned his attention to Africa.[1]

The King's first move was to convene an international geographical conference in Brussels in 1876, where the International African Association was formed to explore the interior of the continent. Two years later, with the help of British, French, German, Dutch, and Belgian financiers, he formed the Comité d'Études du Haut-Congo to investigate trading prospects around the Congo River and the possibility of building a railway around the series of rapids and waterfalls separating the lower Congo River at Matadi from Stanley Pool. The non-Belgian financiers withdrew from the project quite soon, leaving the Comité under Leopold's control; he continued to use its name until 1882 when he introduced the title Association Internationale du Congo. This Association was not at all

[1] Neal Ascherson, *The King Incorporated; Leopold the Second in the Age of Trusts* (Allen & Unwin, London, 1963), *passim.*

international, but served as a cloak for Leopold's manoeuvring to acquire a personal empire that he could eventually turn over to Belgium.[2]

Leopold was a shrewd man, cleverly using humanitarian and commercial interest in Africa to hide his true aim of acquiring territory which could be exploited for Belgium's advantage, and a skilful manipulator of people. It is most unlikely, however, that he could have succeeded in achieving his goal had he not been able to take advantage of some of the great-power rivalries in Africa. Britain was particularly anxious to limit French territorial expansion, while Germany generally distrusted Britain. On 26 February 1884, in an attempt to contain the French, Britain signed a convention recognizing Portuguese claims to the land north and south of the mouth of the Congo River. As the river mouth was the only practical outlet for the trade Leopold wanted to develop in the interior, transit agreements between the Association Internationale du Congo and the Portuguese government would, it seemed, have to be arranged. In April, in order to strengthen his negotiating hand, Leopold offered France the preferential right of acquiring the Association's territories if the Association ever decided to abandon them. Leopold had no intention of giving up the Congo to anyone other than Belgium, but France was not alone at the time in thinking that the heavy financial burden the Congo was already placing on the King might force him to do otherwise.[3] By being able to threaten Portugal with the possibility of the French taking over his territory—France being potentially a more dangerous and predatory neighbour than the Association—Leopold thought he would be in a good position to secure favourable terms. In the event, the Anglo–Portuguese Convention was not ratified because of German opposition. The right of preference did not serve its original tactical purpose, but did prove quite useful none the less. In addition to Britain's wanting to limit French expansion, Germany wanted to keep both Britain and Portugal out of the Congo, while

[2] Ruth Slade, *King Leopold's Congo; Aspects of the Development of Race Relations in the Congo Independent State* (Oxford University Press for the Institute of Race Relations, London, 1962), p. 349.

[3] The one loop-hole in the agreement for the right of preference was closed in Apr. 1887 when France agreed that her right would not apply in the event of a cession of the Association's territory to Belgium. The right was not completely dead, however, as it was confirmed after Belgium took over the Congo in 1908. In Jan. 1960, when discussions of the terms for the Congo's independence were in progress, the French government rather ludicrously informed the Belgians that the right of preference was still valid. The point was not pressed.

both Germany and Britain hoped to profit from the growth of free trade in the Congo which Leopold was prepared to guarantee. The two powers could be relied on to help make any cession of the Association's territory unnecessary. The right of preference undoubtedly also helped smooth negotiations between France and Leopold as they settled the boundaries between their territories.[4]

The United States was the first country to recognize the Association officially, on 22 April 1884. In accepting the right of preference, France implicitly recognized the Association's sovereignty by accepting its right to cede territory. Neither the recognition treaty with the United States nor the exchange of notes embodying the right of preference specified the limits of the Association's territory. The first mention of frontiers appeared in the treaty in which Germany recognized the Association in exchange for guarantees of free trade in its territories and most favoured nation status. Before negotiations for this treaty had begun, Leopold had dreamt of an empire stretching from the mouth of the Congo to the Nile and 'towards the possessions of the Sultan of Zanzibar'. As Germany had her own ideas about the Sultan's lands in East Africa, Leopold was forced to set his easternmost frontier at 30° east longitude, far short of his coastal goal. The southernmost boundary was set at 6° south latitude, well to the north of Katanga.[5]

On 16 December Britain followed Germany's example, recognizing the Association and receiving guarantees of free trade and most favoured nation status. Britain was the only country given the right of consular jurisdiction in the Association's territory, although she never took advantage of it. No mention of boundaries was made,[6] although the British were not happy about the extent of Leopold's claims. They distrusted his intentions regarding the Nile, and feared a possible French take-over. They did not, however, withhold recognition, as the presence of the Association in the area around the

[4] Slade, pp. 35–43; J. Stengers, 'Léopold II et la fixation des frontiers du Congo', *Le Flambeau*, xlvi, no. 3/4 (1963), 153–97; R. S. Thomson, *Fondation de l'État Indépendant du Congo: un chapître de l'histoire du partage de l'Afrique* (Office de Publicité, Brussels, 1933), pp. 164–8; S. E. Crowe, *The Berlin West African Conference, 1884–1885* (Longmans, Green & Co. for the Royal Empire Society, London, 1942), *passim*. A. B. Keith, *The Belgian Congo and the Berlin Act* (Clarendon Press, Oxford, 1919), *passim*. Pierre van Zuylen, *l'Échiquier congolais ou le secret du roi* (Charles Dessart, Brussels, 1959), pp. 82–90 and 135–46.

[5] E. Hertslet, *The Map of Africa by Treaty*, 3 vols. (HMSO, London, 1894, 3rd ed. revised by R. W. Brant, 1909), ii. 572–3.

[6] Ibid. ii. 573–7.

Congo River served the purpose of the Anglo–Portuguese Convention —at least for the time being—of keeping the French out of it.[7]

Leopold then turned his attention to the problem of gaining official French recognition, a task complicated by the fact that the French and the Association made a number of conflicting territorial claims. In accepting the right of preference France had agreed to respect the Association's stations, but had made no formal statement of the frontiers between the lands they claimed. On Christmas Eve 1884 Leopold, having given up some of his claims around and north of Stanley Pool, traced a line on a map with his finger bringing the Kasai and Katanga regions into his domain, more than generous compensation for what he had surrendered. On 5 February 1885 France formally recognized the Association and its extended boundaries in return for the usual guarantees of free trade and most favoured nation status. Although the boundaries between French and Association possessions were set out in detail in the text of the treaty, France merely declared herself ready to recognize the neutrality of the Association within the boundaries shown on an 'annexed map' which, like a map annexed to the treaty with Germany, was not issued with the published text.[8]

The Berlin Conference of 1884/5 established what were supposed to be the ground rules for the European take-over of Africa, although many of its provisions were never enforced. The Conference's final act established the 'Conventional Basin of the Congo' defined as 'all the regions watered by the Congo and its affluents, including Lake Tanganyika, with its eastern tributaries'.[9] Within the Conventional Basin, the provisions of the act—neutrality, free trade, arbitration rights, etc.—were to operate. There was no definition of the boundaries between the parts of the Conventional Basin which might eventually be occupied by different countries. The question was discussed during the Conference, but the delegates seem to have wanted to avoid making any official pronouncements which would lead to the immediate extension of the 'scramble for Africa' into the interior.[10] This was a futile position as the scramble had already been

[7] R. R. Robinson and J. A. Gallagher with Alice Denny, *Africa and the Victorians* (Macmillan, London, 1961), pp. 169–74. Stengers, 'Léopold II et la fixation . . .', pp. 190–1; Crowe, pp. 16–33 and 142–51; Van Zuylen, pp. 67–73.

[8] Hertslet, ii. 564–656.

[9] 'Chapter 1 of the General Act of the Conference of Berlin' in ibid. ii. 71.

[10] 'Rapport de la commission chargée d'examiner le projet de déclaration relative aux occupations nouvelles sur les côtes d'Afrique', annexe No. 1 to Protocol in C. 4361, Africa No. 4 (1885), pp. 200–6.

extended, but any other attitude would probably have made it impossible for the delegates to agree on the terms of the Final Act.

In editing the Conference proceedings for publication, the Germans annexed to one of the protocols a map showing the frontiers within the Conventional Basin as established by the treaties with France, Germany, and several other countries. As the map had not been submitted to the Conference delegates, it did not form an official part of the 'Berlin Act'. No objections to what it showed were raised, however, and Leopold proceeded to form the Independent Congo State, and became its King-Sovereign. To consolidate his position he had to secure formal acceptance of the extensive boundaries he claimed by the countries that had signed the Berlin Act. This he did with what has been called one of his 'plus beaux coups d'audace'.[11]

On 1 August 1885 the Independent Congo State announced its formation, and issued a circular declaring its neutrality. The really important part of the declaration was the specification of the limits within which neutrality was to operate—effectively the frontiers of the Congo State, which corresponded to the frontiers accepted by France.[12]

France clearly had no reason to raise any objections to the boundaries, as she still had the hope of acquiring the entire State. Germany might have objected as these boundaries were considerably more extensive than those she had officially recognized earlier. The extensions were not into the territory in which Germany was specifically interested, however, and Leopold had in any case anticipated such an objection by sending a copy of the draft declaration to Bismarck the previous June, securing the Chancellor's prior approval.[13] When the Declaration of Neutrality was formally issued, Bismarck acknowledged its receipt noting the discrepancy in the boundaries, but this only to show that it had not gone unnoticed. He had no reservations in fully and formally recognizing the State's claimed frontiers.[14]

It was only Britain's reaction about which Leopold was uncertain.

[11] J. Stengers's article on Edmund van Eetvelde in *Biographie coloniale belge*, iii (IRCB, Brussels, 1951), col. 329.

[12] 'Memorandum on recognizing the . . . boundaries of the Congo State', 15 May 1891, prepared by Lord Vivian, British Minister in Brussels, 1884–91, FO 84/2118. Thomson, pp. 300–3.

[13] Ibid., p. 301.

[14] Vivian's Memorandum, op. cit., and minute on drafts of Salisbury to Vivian No. 9, Africa, 19 Sept. 1891, FO 84/2117.

In view of the British attitude towards his extensive claims the previous year, they might have been expected to refuse to recognize the boundaries set out in the Declaration. The opportunity to protest was lost, however, when the assistant clerk, in charge of the African Department while more senior officials were on holiday, acting on the assumption that the frontiers had been embodied in the Berlin Act, drafted a reply simply acknowledging receipt of the declaration. Sir Percy Anderson, Under-Secretary in charge of the African Department, was several years later to call this a 'stupid blunder', but too late to raise any formal protest.[15] There is no certainty that any protest would have been made as Leopold's extended claims did not bring him any nearer to the Nile which was Britain's primary concern. There was no British interest in the southern portions of the Independent Congo State until some years later. Leopold was free to focus his attention on building the railway from Matadi to Stanley Pool, and other more immediate problems.

RHODES THREATENS

In 1806 two Portuguese–African slave traders (*pombeiros*) reported on a trip they had made overland from Angola to Moçambique, in the course of which they passed through Katanga where they found copper used as money and ornamentation. This report was later confirmed by Livingstone and V. L. Cameron, the explorer who became Leopold's associate in the development of Katanga, and by the Plymouth Brethren missionaries, Frederick Stanley Arnot and Dan Crawford who lived in Katanga for many years.[16] None of these reports provided any details of the extent and value of copper deposits, and although Leopold probably knew that copper was to be found in Katanga, it had played no part in his decision to include the region in his personal empire.

The King-Sovereign remained quite indifferent to Katanga until 15 October 1889 when he saw in *The Times*, which he read regularly and avidly, an article on the charter it was proposed to give Cecil

[15] Stengers, 'Léopold II et la fixation . . .', pp. 188–90.

[16] A. Verbeken and M. Walraet, *La Première Traversée du Katanga en 1806; voyage des 'pombeiros' d'Angola aux Rios de Sena* (IRCB, Brussels, 1953), pp. 75 and 96; V. L. Cameron, *Across Africa* (Baldy Isbister & Co., London, 1877), i. 318–20, and ii. 149 and 329; Dan Crawford, *Thinking Black; Twenty-Two Years without a Break in the Long Grass of Central Africa* (Morgan & Scott, London, 1912), p. 164; R. Rotberg, 'Plymouth Brethren and the Occupation of Katanga', *Journal of African History*, v (1964), 285.

Rhodes and the British South Africa Company (Chartered). He immediately contacted Albert Thys, one of his staunchest financial supporters, who had founded the Compagnie du Congo pour le Commerce et l'Industrie (CCCI) which was involved in most of the development that had already taken place in the Congo, saying that they must not delay sending an expedition into Katanga.[17] He then informed the British Minister, Lord Vivian, that he wanted to make immediate contact with the company in order 'to lose no time in opening . . . friendly relations with' them. He particularly wanted to know the northern limit of Chartered's territory.[18] After seeing a copy of the Charter, which did not specify a northern frontier, he and the company exchanged mutual assurances of their hopes that under Lord Salisbury's auspices they would maintain good relations.[19]

Pious words needed the support of firm action. In February 1890 George Cawston, London Manager of the British South Africa Company, went to Brussels to assure the international slave-trade conference meeting there of the company's support for the suppression of the trade in their own and neighbouring territories.[20] Lord Vivian presented Cawston to Leopold who, after listening to an explanation of Chartered's plans for developing its territory, reiterated his desire for amicable relations between the Congo State and the Company, adding the hope that their possessions might eventually become co-terminous. He went on to suggest that an agreement for mutual support should be drawn up, defining their frontiers as far as possible and giving the Congo State recruiting rights in Chartered country. Cawston agreed to bring these suggestions to the attention of the company's Board of Directors. Leopold was clearly anxious about Chartered's intentions and wanted to contain them as quickly as possible.

Cawston, Rhodes, and the Chartered directors agreed that some control over Katanga's mineral resources would be a valuable asset for the company.[21] There was considerable disagreement, however, as to the possible means of securing that control. Rhodes took the extreme view that Leopold could not dispose of Katanga as it was

[17] Leopold to Thys, 16 Oct. 1889, Thys Papers, Vol, xxii, No. 58.

[18] Vivian to Salisbury, No. 110, Africa, 19 Oct. 1889, FO 84/1946.

[19] Ibid., No. 125, Africa, 9 Nov. 1889, FO 84/1946 and Salisbury to Vivian, No. 162, Africa, 18 Dec. 1889, FO 84/1944.

[20] Ibid., No. 6 Africa, confidential, 8 Feb. 1890, enclosing Cawston to Vivian, 6 Feb. 1890, FO 84/2004.

[21] Cawston to Rhodes, 14 Mar. 1890, Cawston Papers, Vol. ii.

not really his; he had not fulfilled the Berlin Act requirement of 'effective occupation'. Katanga was still a 'no-man's-land' that anyone who could establish himself there could take. Cawston took the opposite position, maintaining that Katanga unquestionably belonged to the Congo State, but that Leopold was tired of the heavy financial burden the State was placing on him, and would be prepared to co-operate with Chartered rather than seek German assistance. Even if Leopold did not want to sell Katanga outright, he and Chartered could agree on a profit-sharing scheme of some kind. Cawston was somewhat uncertain about the effect France's right of preference might have on any such arrangement, but had no doubt that Katanga belonged to Leopold who could dispose of it in any way he wished. Chartered's best course was to reach an understanding with the King.[22]

Leopold was by now convinced that Katanga contained important mineral resources, including gold, which had to be prevented from falling into foreign hands.[23] He pressed Chartered to agree to fix their common boundary and to accept his suggestion that the line of 4° south latitude could serve the purpose. Chartered preferred to await the outcome of the negotiations they were carrying out for the settlement of their frontier with German East Africa. Much depended on whether or not they would have access to Lake Tanganyika. Leopold then approached Salisbury, proposing a more definite frontier, and offering the company a strip of land five miles wide between Lakes Tanganyika and Bangweulu.[24] The King gave the impression that the company had already agreed to this in principle, but they had not. In May when Leopold was in England privately (supposedly incognito but fooling no one) to attend the unveiling of an equestrian statue of Prince Albert in Smith's Lawn, Windsor Park, to which Queen Victoria had not invited him, he put this proposal to Chartered, wanting in return a strip of land linking the Congo to a port on Lake Nyassa as well as recruiting rights in Chartered territory. The company only even considered this because their negotiations with Germany seemed to be going badly and it looked as if they could only have access to Lake Tanganyika by

[22] Cawston to Rhodes, 6 Feb. and 14 Mar. 1890, Cawston Papers, Vol. ii. Rhodes and Leopold did in fact discuss the possibility of joint operations in Katanga early in 1899. (Note of interview with Rhodes, by Leopold, 3 Feb. 1899, in Leopold's Papers.)

[23] Leopold to Thys, 14 Feb. 1890, Thys Papers, Vol. xxiv, No. 12.

[24] Van Zuylen, pp. 202–4 and notes 1 and 2.

accepting the land Leopold was offering them. Otherwise there was no reason to allow Leopold to extend the Congo State at Chartered's expense.[25] In fact, the treaty in which the southern border of German East Africa was settled, along with other outstanding questions between Britain and Germany, gave Chartered access to the Lake, and Leopold's plans came to nothing.[26]

The King was very much perturbed by articles appearing in the British press supporting Rhodes's view of Katanga. He was particularly upset by a map of central Africa in the *Graphic*, a weekly magazine published in London, showing large parts of the Congo State as belonging to Chartered. The royal anger was directed primarily against Mr. (later Sir) Harry Hamilton Johnston who had provided the map and who as Commissioner and Consul General for the British Central African Protectorate (Nyasaland) and, at the same time, administrator of Chartered's north-eastern region had openly made treaties with chiefs in areas belonging to the Congo.[27] The King pressed for an agreement with Chartered that would put an end to Johnston's predations[28] and sought assurances that any treaties Johnston made in the Congo would be disavowed by the British government.[29]

Johnston was if anything more firmly convinced than Rhodes that Katanga should come under British control. Before going out to take up his post in central Africa, he told Rhodes that Cawston's view of the Congo's boundary was 'maddeningly conciliatory' and his 'indifferent tolerance of Belgian ambitions . . . irritating'.[30] Johnston expanded his theme to Lord Salisbury, saying that he had 'reason to believe that the King of the Belgians wants to chase us out of an important piece of territory which we need to complete the new province of British Central Africa . . . Katanga . . . The King of the Belgians has no right to it.' Johnston believed, for reasons best known to himself, that it was only through the 'indulgent carelessness' of British map-makers that Katanga was included in the Congo State.

[25] Cawston to Rhodes, 20 June 1890, Cawston Papers, Vol. ii.

[26] Anglo–German Agreement of 1 July 1890, in Hertslet, iii. 899–904.

[27] The events leading to the establishment of this rather complicated dual position, Johnston's fulfilling of his often conflicting responsibilities, and his relations with Rhodes are discussed by Roland Oliver in *Sir Harry Johnston and the Scramble for Africa* (Chatto & Windus, London, 1957), pp. 124–244.

[28] Vivian to Salisbury, No. 42 Africa, confidential, 12 July 1890, FO 84/2024 and Leopold to Salisbury, 12 July 1890, FO 84/2086.

[29] Ibid., No. 45 Africa, very confidential, 22 July 1890, FO 84/2024.

[30] Johnston to Rhodes, private, 25 July 1890, Rhodes Papers, Vol. iiiA.

He did not see why Leopold should be treated by the British in any more generous or sentimental a way than France, Germany, or Portugal. In describing Leopold himself, Johnston was even more forthright:

> From the Belgian point of view he is an admirable patriot, who has by many a hook and crook, by many a wile and intrigue, by much expenditure of his own and not a little of that subscribed by the rich English people whom he bamboozled, created a fine African Empire for little Belgium. But why his enterprise should be viewed by *us* . . . with indulgence . . . I cannot conceive.[31]

Cawston, somewhat more closely in touch with Leopold and probably more aware of the practical possibilities of working with him, remained convinced that there was no chance that Chartered could ever annex Katanga. He saw Johnston's actions as the cause of considerable harm because they aroused suspicion which would only serve to spur Leopold on to send expeditions to Katanga. Salisbury did not, in any case, agree with Johnston, and sent the King a conciliatory reply.[32]

Rhodes was wrong in saying that the Berlin Act required Leopold to occupy Katanga effectively to give him a firm claim to it, as the Act referred only to the effective occupation of coastal areas, most of which had already been claimed and occupied by the time the Conference was called. There was ample other historical justification for Rhodes's point of view, however, and whatever the position in law, it was clear both to him and to Leopold that whichever of them managed to establish himself in Katanga first would be in a strong position. If Chartered actually secured a concession from or the submission of M'siri, the autocratic overlord of the area, Leopold would be at a serious disadvantage in negotiating with them. On the other hand, if Leopold secured a foothold in Katanga first, Chartered would lose any reasonable claim they might ever have had to include Katanga within the territory under their administration, and could only profit from Katanga's potential wealth by co-operating with the King. Neither could lose any time in making contact with M'siri.

In March 1891 Joseph Thomson, a well-known explorer in Africa who had travelled extensively in East Africa and Nigeria, left

[31] Johnston to Salisbury, 25 Aug. 1890, cited in Oliver, p. 194 n.

[32] Cawston to Rhodes, part confidential, 3 Sept. 1890, Rhodes Papers, Vol. iiiA, and Oliver, pp. 194–5.

England for Cape Town, to discuss with Rhodes details of an expedition he was to lead into Chartered country and Katanga.[33] Pressed and supported by the King, Thys began to arrange for a Belgian expedition as well. While this was being organized, at the end of 1890, an expedition under Lieutenant Paul Le Marinel, Commander of the Congo State's Lualaba District, was sent to establish a post in Katanga, and if possible to secure M'siri's submission to the State's authority. Le Marinel arrived at Bunkeya, M'siri's capital, in April 1891 and after some difficulty was permitted to establish a post at Lofoi, about 30 kilometres to the east, near the present village of Mukebo. He was unable to carry out the rest of his instructions. After leaving Bunkeya he received a letter from M'siri, but this merely indicated the chief's interest in co-operating with Le Marinel to secure help in putting down internal unrest which had been increasing in his domain in response to his despotic rule. The letter did not constitute an act of submission, and M'siri subsequently disavowed it.[34]

Thys organized an expedition under the leadership of Alexander Delcommune, who had joined the CCCI in 1886, and led an expedition to the Haut-Congo in 1888. Delcommune left Europe in 1890 for Boma, to travel first to Stanley Pool, then to Katanga along the Congo, Kasai, and Sankuru Rivers.[35] Before the expedition got very far, its sponsorship changed as a new company was formed and took it over from the CCCI. In early December 1890 Thys passed on to Leopold a proposition made by Albert Ochs, a London merchant with connections in South Africa, for the establishment of a company similar to Chartered to develop Katanga. Ochs had discussed this scheme with financiers in London and Germany, and claimed to have considerable support. After preliminary discussion with Samson Wiener, Leopold's confidential lawyer who acted as his intermediary in this and other business affairs, Ochs changed his approach slightly and proposed the formation of an international company with one million pounds capital. The company would be given a concession to

[33] Lord Abercorn, President of Chartered, to Cawston, 29 Mar. 1890, Cawston Papers, Vol. ii. See also J. B. Thomson, *Joseph Thomson, African Explorer* (Sampson Lowe & Marston, London, 1896), and Robert I. Rotberg, *Joseph Thomson and the Exploration of Africa* (Chatto & Windus, London, 1971).

[34] A. Verbeken, *M'siri, roi du Garenganze* (Cuypers, Brussels, 1956), pp. 181–94. E. Verdick, *Les Premiers Jours au Katanga* (CSK, Brussels, 1952), pp. 17–34.

[35] Verbeken, *M'siri* . . ., pp. 195–208; Slade, *King Leopold's Congo*, pp. 130–1; Verdick, pp. 54–5.

exploit mines, construct railways, and provide necessary roads and navigation services. This, Ochs claimed, would put a stop to Chartered's 'empiètements', but Thys was very wary, indeed suspicious. He made inquiries into Ochs's background in South Africa, and was not entirely pleased to find that Ochs had been involved with Rhodes and Chartered, but had nearly been compromised in dealing with Argentine shares. Thys assured the King that he would be very careful in listening to what Ochs had to say.[36]

Ochs in fact had very little to do with Chartered. His group of companies had a variety of interests, some in the Transvaal, and some in Moçambique. Thys ultimately lost his distrust of Ochs as the two men became colleagues in the Companhia da Moçambique, one of the chartered companies responsible for the administration and development of Moçambique for the Portuguese. Leopold was not unwilling to establish a company along the lines Ochs suggested, and by the end of December 1890 fairly detailed plans were going ahead for the formation of a company in which Chartered was to have a large share—eight out of twenty million francs authorized capital. French, Belgian, and Portuguese interests were to take up the remaining capital, but no single group other than Chartered was to have more than four million francs.[37] Chartered did not have the control of Katanga she wanted, and when the Compagnie du Katanga was formed on 15 April 1891, she did not participate in it at all. British interests were represented primarily by V. L. Cameron, and by Ochs and some of his companies. Maurice Bunau-Varilla, a Parisian newspaper proprietor, and Seligman Frères & Cie were the principle French participants, while the CCCI, Léon Lambert, a banker who often supported the King's schemes, and several other Belgian banks and companies supplied the Belgian contribution. Leopold was able to keep his eye on what happened within the company through several directors, Baron Constant Goffinet who was Administrator of the Civil List, Count John d'Oultrement, Grand Marshal of the Court, and Thys. The Compagnie du Katanga took over from the CCCI all responsibility for opening up Katanga, and getting there before Chartered.

In an attempt to ensure that even if Thomson were successful in securing M'siri's submission, Chartered's claim to Katanga would not

[36] Thys to Leopold, 4 and 6 Dec. 1890, Leopold II Papers Nos. 106/52 and 62.

[37] Tiano, one of Cawston's French associates, to Cawston, 23 Dec. 1890, Cawston Papers, Vol. iii.

be recognized, Leopold early in 1891 asked the Foreign Office for an assurance that the Congo State's boundaries as stated in the 1885 Declaration of Neutrality would be respected.[38] The impression Leopold gave of wanting to avoid further expenditure in Katanga if he could is somewhat belied, although not necessarily contradicted by the assurances he had earlier given Thys that the State would meet any part of the cost of Delcommune's expedition that the company could not.[39] Sir Percy Anderson, who probably took Johnston less seriously than Salisbury did, pointed out that although Johnston had long coveted Katanga, none of the signatories to the Berlin Act had in five years raised any objection to the map annexed to the Conference protocols, and that therefore the boundaries shown on the map had essentially been recognized.[40] However stupid a blunder had been committed in acknowledging receipt of the Declaration of Neutrality, nothing could now be done about it. Anderson drafted a reply to Leopold, accepting the Declaration as the Congo's statement of the boundaries, but adding that this did not necessarily coincide with the British version of them.[41] Salisbury did not approve Anderson's draft, and chose to cloud the issue by referring to the declaration of neutrality as mentioned in the Berlin Act. Stressing that this contained no definition of territorial limits, he assured Leopold that Chartered would be required to respect the State's rights as indicated in the Act.[42] This wonderful obfuscation must have made Leopold furious. In polite, formal language he corrected Salisbury, and renewed the assurance, which by this time may well have sounded rather hollow, of his desire for friendly relations with Chartered who, he maintained, had no reason to be jealous of the Congo State in Katanga.[43]

Leopold, however, did have reason to be concerned about Chartered, as was further confirmed in April 1891 when Chartered informed him that 'having regard to the friendly relations existing between [them] and the Congo State, [they] thought it only right' at once to inform him that they had sent Thomson to M'siri's country and that according to cabled reports, Thomson had already signed a

[38] Vivian to Salisbury, No. 2 Africa, very confidential, 3 Jan. 1891, and minute thereon by Anderson, FO 84/2118.
[39] Leopold to Thys, 5 July 1890, Thys Papers, Vol. xxvii, No. 26.
[40] Memorandum by Anderson, 15 Jan. 1891, FO 84/2117.
[41] Ibid.
[42] Salisbury to Vivian, No. 9 Africa, 19 Jan. 1891, FO 84/2118.
[43] Vivian to Salisbury, No. 7 Africa, 21 Feb. 1891, FO 84/2118.

treaty with that chief. Not surprisingly, the Congo State registered a strong formal protest.[44] This letter from Chartered might seem rather odd, particularly in the light of the publication, just prior to this time, of a map in which Cawston confirmed his conviction that Katanga was part of the Congo State. Cawston claimed, however, that as M'siri's jurisdiction extended beyond the State's boundaries it was perfectly possible for Thomson to have made a treaty without trespassing. Chartered had informed Leopold of what had happened, according to Cawston, because concealment of the information while the company and the State were still on friendly terms 'might be taken as an underhand proceeding'.[45]

In fact there was no reason for anyone to be particularly bothered. Thomson had not reached Bunkeya at all. He was an energetic man, but very stubborn, thoroughly convinced that he always knew what he was doing, and needed no one else's advice. He was also very quick and imprecise in his work. At Kimberley, where he and Rhodes were joined by Johnston to plan his expedition, he refused to accept any suggestions as to the best route to follow. He insisted on approaching Katanga from the south-east, and was in the swamps south of Lake Bangweulu, some 200 miles short of his objective when his remaining porters—many had already died of smallpox—threatened to desert rather than go any further. Thomson himself was suffering from bilharzia and cystitis, and had to turn back. His failure was the more ignominious as he gathered no geographical information that could be of any use to Rhodes or anyone else. He did sign a number of treaties with local chiefs—including one of M'siri's vassals who had adopted his overlord's name—or people he thought were chiefs, which later served as the basis of Chartered's claims to mineral rights in Northern Rhodesia. They were 'absurdly worded' and had no legal validity. Johnston issued certificates of claim based on them, but they were never accepted by the Foreign Office. Thomson's trip was to all practical intents and purposes a waste of time, effort, and money.[46]

At Kimberley, Rhodes and Johnston had been preparing another

[44] Van Eetvelde to Vivian, 13 Apr. 1891, enclosing copy BSAC to Comte Jean d'Oultremont, Maréchal de la Cour, 9 Apr. 1891, FO 123/293.

[45] Minutes on Vivian to Salisbury, No. 30 Africa, 14 Apr. 1891, FO 84/2118.

[46] Thomson, *Joseph Thomson*, pp. 247–69; Rotberg, *Joseph Thomson . . .*, pp. 256–89 and 298–306. Oliver, p. 222. Joseph Thomson, 'To Lake Bangweolo and the Unexplored Regions of British Central Africa', *Geographical Journal*, i (1893), 97–121 and 192.

string for their bow. Johnston, acting on Rhodes's authority, arranged for Mr. (later Sir) Alfred Sharpe, then on a hunting trip in southern Africa, later Vice-Consul in Nyasaland, to travel to M'siri's capital by way of Lakes Tanganyika and Mweru. Except for some difficulties with Chief Kazembe north of Lake Mweru, Sharpe had a relatively easy journey, but arrived at Bunkeya looking rather less resplendent than might have been desirable for making a good impression on the chief. Sharpe was able to have several conversations with M'siri, but when Thomson failed to arrive to support his story, Sharpe found the chief less amenable. M'siri refused to sign any treaty, in part because the missionary Arnot, who was away from Bunkeya at the time, had strongly advised him not to do so; in part because he was afraid of anyone approaching his kingdom from the east, as Sharpe had done. Sharpe left rather quickly.

Delcommune's expedition, which had left Europe at about the same time as Le Marinel set out from Lusambo arrived in Bunkeya not long afterwards. Delcommune's instructions were to support Le Marinel, attempt to secure M'siri's submission in the event of Le Marinel's failure to do so, and generally explore the area. M'siri also tried to use Delcommune to help quell opposition to his rule, but was not prepared to surrender any of his formal authority. Delcommune was no more successful than those who had preceded him.[47]

M'siri had established himself in Katanga in about 1856 to carry on trade that his father had started. He was from the Sumbwa tribe living in what is now north-western Tanzania, and over the years built up trade in copper, ivory, salt, and slaves. He also established himself as ruler of an increasingly large territory, which by the late 1880s was about the size of Great Britain. His rule was despotic, and aroused strong opposition. By the time Europeans began to be interested in Katanga, his hold over his subjects was becoming weaker as they rebelled against his harsh rule. Particularly troublesome for him at this time were the Basanga, a group living in northern Katanga with whom M'siri had originally allied himself, only to subjugate them later. Had M'siri been able to face the expeditions sent by Leopold and Rhodes with a strong, united people behind him, the

[47] A. J. Hanna, *The Beginnings of Nyasaland and North-Eastern Rhodesia, 1859–1895* (Clarendon Press, Oxford, 1956), p. 160; J. G. Lockhart and C. M. Woodhouse, *Cecil Rhodes: The Colossus of Southern Africa* (Hodder and Stoughton, London, 1963), pp. 242–3; Slade, p. 132; Alfred Sharpe, 'A Journey to Garenganze', *Proceedings of the Royal Geographical Society*, N.S. xiv (1892), 36.

task of securing his submission would have been much more difficult than it turned out to be.[48]

The Basanga revolt had been going on for about nine months when a second expedition sponsored by the Compagnie du Katanga reached Bunkeya. This was under the leadership of Captain W. G. Stairs, a Nova Scotian in the British army who had been with Stanley in the Congo between 1887 and 1889 on the expedition to relieve Emin Pasha. By the time Stairs arrived, M'siri's effective authority was said to have dwindled to cover a radius of only about fourteen miles from his capital. Three days later Stairs had an inconclusive interview with M'siri, during which the chief asked for help not only in crushing the Basanga but also in getting rid of Lt. Legat, whom Le Marinel had left in charge of the post at Lofoi. M'siri was still trying to play Europeans off against one another; he had previously gone so far as to suggest to Legat that he would leave the country to the Basanga if he could find a safe refuge east of Lake Mweru. A second interview with M'siri two days later brought no better results. That evening Stairs decided to put an end to M'siri's havering about accepting Congo State authority and had one of his men hoist the State's flag on the hill outside the village. Surprisingly, this produced no apparent reaction. The next morning it was discovered that M'siri had fled during the night to the village of one of his wives. Captain Bodson, a Belgian officer who was second-in-command of the expedition, went to remonstrate with the chief and found him surrounded by bodyguards. Bodson chided M'siri for not continuing the negotiations he had begun with Stairs. M'siri threatened Bodson with a sword, Bodson fired two bullets into the chief, his companion, the Marquis de Bonchamps, adding a third. Almost immediately Bodson was felled by a bullet in his side. M'siri was dead; Bodson succumbed several days later. The way was clear for Stairs to arrange the succession of chiefs in such a way as to ensure that there could be no further effective opposition to rule by the Congo State.[49]

The ravages of war and the famine resulting from the destruction of

[48] A very full account of M'siri's life, his methods of rule and trade, his contacts with Europeans, and his eventual downfall is given in Verbeken, *M'siri*. His role in East–West trade is discussed in Alison Smith, 'The Southern Section of the Interior, 1840–84' in R. Oliver and G. Mathew (eds.), *History of East Africa*, i (Clarendon Press, Oxford, 1963), 276.

[49] 'Le Journal du Capitaine Stairs' in *Le Congo Illustré*, ii (1893), especially those parts appearing in No. 25 of 3 Dec., pp. 197–9. Verbeken, pp. 209–43. J. A. Moloney, *With Captain Stairs to Katanga* (Sampson Lowe & Marston, London, 1893), pp. 182 ff.

crops and people's inability to replant them also affected the members of the expedition who were very short of food. Stairs wanted to leave immediately, all the more so as his health was deteriorating rapidly. He had to wait, however, until another expedition under the leadership of Lucien Bia, who had served as an officer of the Congo State for several years in the area around the mouth of the Congo and north of Stanley Pool, reached Bunkeya. Once Bia arrived, Stairs lost no time in setting out for the coast, but died at Vicente on 9 June 1893.[50] Bia took over the further exploration and pacification of Katanga. The most important aspect of the exploration work was done by the expedition's geologist, Jules Cornet, who travelled through much of southern Katanga and found a number of copper mines. As he had only very limited time and equipment at his disposal, it is surprising that he felt justified in concluding that the copper deposits were not sufficiently rich to justify the great expense which would be involved in developing them, not only because of the relatively low concentration of copper but also because of the distance of the mines from any cheap form of transport.[51]

Following the receipt of Cornet's report the Compagnie du Katanga decided not to proceed with the development of the mines and turned their attention to the Lomami region. Since then there has grown in Belgium what amounts almost to a cult which sees Cornet as the father of the Katanga copper industry. Cornet himself did nothing to disillusion people on this point. Ten years after his initial work, when it was proved that the copper deposits were in fact very rich and well worth considerable expenditure to exploit them, he wrote that he had been right in attributing great importance to the deposits.[52]

Cornet had been unable to do more than a superficial survey. A more fully equipped expedition visited Katanga under the leadership of Captain Charles Lemaire in 1898, by which time rumours of mineral wealth in Katanga had grown considerably. Lemaire reported, however: 'The mineral treasures which have been for a long

[50] Moloney, p. 276.

[51] R.-J. Cornet, *Katanga: Le Katanga avant les Belges et l'expedition Bia-Francqui-Cornet*, 3rd ed. (Cuypers, Brussels, 1946), *passim.* Jules Cornet, 'Les gisements métallifères du Katanga', first published as a *mémoire* of the *Société des Sciences des Arts et des Lettres du Hainaut*, 5th Ser. viii (1894), 3, subsequently in *Revue Universelle des Mines*, 3rd Ser. xxvii (1894), 217, and later in *Bullétin de la Société Belge de Géologie, de Paléontologie et d'Hydrologie*, xvii (1903), 3.

[52] In *Bulletin de la Société Belge de Géologie*, xvi (1902), 651.

time so liberally ascribed to Ka-Tanga [*sic*] did not reveal themselves to us.'[53] Neither Cornet nor Lemaire appears to have taken sufficient account of the fact that the surface deposits had been worked by Africans for a long time, or to have examined deposits carefully at any depth. It was only when this was done by George Grey, Franz Studt, Michael Holland, and Henri Buttgenbach that the true value of Katanga's copper was appreciated.

With M'siri dead and the succession of compliant rulers secured, the Independent Congo State's claim to Katanga was clearly established, even by Rhodes's criterion. Leopold had indeed 'got there first'. The final acceptance of this fact was embodied in a treaty signed on 12 May 1894 by Great Britain and the Congo State, setting the frontier between the State and Chartered's northern territory. With the exception of some changes brought about by improved knowledge of the region and some negotiated alterations, the frontiers were the same as those set down in the 1885 Declaration of Neutrality. The treaty also established the boundary between British and Congo State territory north of German East Africa. Britain also gave Leopold leases on land around Lake Albert, including the port of Mahagi, the Bahr-el-Gazal province of the Sudan, and on an enclave at Lado on the Nile. The lease on Lado was for Leopold's lifetime only, while the others were to pass to his successors as long as the Congo remained an independent state or a Belgian colony. The leases served British interests by putting Leopold into the southern Sudan as an obstacle to France, and seemed to further Leopold's desire to expand to the north-east and have a link to the Nile. French opposition was so strong, however that Leopold could never establish himself in the Bahr-el-Gazal, and the lease was cancelled in 1906. A third provision of the treaty suffered a more rapid demise. The Congo State granted Britain a lease on a twenty-kilometre strip of land west of Lake Tanganyika connecting British territory north and south of German East Africa. This might have served as a route for the mythical Cape-to-Cairo Railway or, more immediately and more possibly, for a telegraph line. Germany was so strongly opposed to this lease, however, that Britain and the Congo State agreed to cancel it in June 1894.[54] Leopold had Katanga, but was not immediately interested in doing anything with it.

[53] Charles Lemaire, 'On the Congo: The Belgian Scientific Expedition to Ka-Tanga', *Scottish Geographical Magazine*, xvii (1901), 544.

[54] Robinson and Gallagher, pp. 339–78; Van Zuylen, pp. 231–57.

A CONCESSION FOR LADY WARWICK

The original impetus for the third and ultimately successful assault on Katanga's mineral wealth apparently came from a house party at Easton Lodge, the Essex home of the Countess of Warwick in 1895. In the course of conversation, the Countess told Rhodes that she wanted him to give her a concession in Chartered territory. Rhodes replied that this would only be possible in the area north of the Zambezi River. Building on her 'long friendship' with the Prince of Wales,[55] Lady Warwick was able to rally strong support for her request. After Rhodes had returned to South Africa he received a letter from the Prince asking him to grant a concession to Tyndale White, also a close friend of the Countess and director of Warwick Estates Ltd., who headed a syndicate in which the Countess was the major participant.[56]

For assistance in meeting the royal request Rhodes turned to one of his associates, Robert Williams, an Aberdonian who had first gone to South Africa in 1881 as an engineer for Blaikie Brothers, a Scottish firm building a bridge at Kimberley. After returning to Scotland and trying to settle into work at his family's whisky distillery, he undertook some formal engineering training before returning to Africa, where he soon acquired a good reputation as a mining engineer. In 1891 he and Rhodes formed the Zambesia Exploring Company to open up some part of Chartered territory in what became Southern Rhodesia. Tyndale White was the company's chairman, Williams its

[55] J. F. H. St. C. Erskine, 5th Earl of Rosslyn, *My Gamble with Life* (Cassell, London, 1928), pp. 263–9.

[56] F. Swann, 'My Work in Africa', unpublished typescript of Williams's memoirs, Chapter VII, p. 2. See bibliography for a discussion of the different versions of these memoirs. The Prince's letter itself has not come to light, but there is ample evidence of its general tenor. In 1910 Robert Williams, in expressing his sympathy on the death of King Edward VII, wrote: '. . . it was entirely as the result of a letter which His Late Majesty wrote to . . . Rhodes that . . . Rhodes asked me to advise him on certain matters as a result of which I formed the Tanganyika Concessions Company'. (Williams to Lt.-Col. Sir Arthur Bigge, Equerry to King George V, 23 May 1910, Williams Papers.) Some years later, Sir J. Ulick Alexander, who became a close associate of Williams and director of some of his companies after having served in the Sudan as aide-de-camp to General Sir Francis Reginald Wingate, and later in South Africa as Imperial Secretary, in advising Williams on the publication of his memoirs said: '. . . I feel the whole truth should come out with the words "High Authority" . . . in the place of E. This would leave the C[hartered] Company to explain who the "High Authority" was and why they gave the concession'. (Alexander to Williams, 13 Sept. 1933, Williams Papers.)

African Manager, and his brother Oliver London Manager. It was Oliver Williams who brought the Prince of Wales's letter to Rhodes, although Rhodes had already been told by Alfred Beit of its contents.[57]

After considerable discussion Chartered agreed to give White a 6,000-square-mile concession north of the Zambezi. Williams agreed to be African Manager of the company that would be formed to exploit the concession on condition that the syndicate do nothing to float the company until a satisfactory report had been received on the potential of the area in which the concession was to be located.[58] This appears to have been acceptable to all concerned; White assured Rhodes: 'HRH will be much gratified at the truly liberal spirit in which the recommendation contained in his letter has been met.'[59] To carry out the preliminary investigation, Williams sent Cecil Livingstone Learmonth, whom he had recently met aboard ship from England, to look over the land between the Zambezi and Loanga Rivers and at the same time to locate a 300-square-mile concession granted to Williams personally.[60]

Before Learmonth reported, Williams received a cable from London informing him that the syndicate had floated a company capitalized at £1,000,000 to work the concession. He refused to manage the company until he knew more about the concession area.[61] An exhortation to the effect that the flotation would fail if he did not accept the management failed to move him.[62] He could see no reason why he and his associates in the mining consultants firm, Robert Williams & Co., should risk their reputations in supporting the proposed company without any real knowledge of its assets. The sole object of the flotation appeared to him to be to make money for other people. If there was anything of value in the area his own concession would certainly cover some of it, and he could secure further grants if that seemed advisable.[63] From a telegram saying that he had 'knocked the flotation' he concluded that the concession itself was valued at 'nil' and that *he* was 'valued by the public at £1,000,000'. If this were the case he saw no reason why he and his

[57] Undated memorandum by Williams, Williams Papers.

[58] Col. Charles McKinnon's draft of Williams's memoirs, Chapter VIII. pp. 8–9.

[59] White to Rhodes, 26 July 1895, Rhodes Papers, Vol. iiiB.

[60] Dawson's version of Williams's memoirs, Chapter VII, pp. 7–8.

[61] Williams to Robert Williams & Co., telegram, undated, but probably just prior to 10 Oct. 1895, Williams Papers.

[62] Williams to Dr. L. S. Jameson, 10 Oct. 1895. Williams Papers.

[63] Williams to Robert Williams & Co., 9 Oct. 1895, Williams Papers.

colleagues should jeopardize their reputations for the sake of 'making £300,000 for HRH & Co.'; if 'HRH & Co.' were so anxious for the company to be formed, they should take all the risks.[64] Williams did relent so far as to say that if the company were floated at £500,000 with £250,000 working capital, he would accept the managing directorship. Such a company, he thought, could be made profitable even if the mineral deposits in the area proved valueless.[65] A tentative arrangement was made,[66] but Learmonth put an end to it when he reported, at about the same time as the Jameson Raid occurred, that there was nothing of any value whatever in the proposed concession area.[67] The Raid and the subsequent uncertainty about the security of British investment in South Africa, particularly the question of Chartered's continued existence, made it impossible to proceed with any scheme such as that envisaged by Lady Warwick's syndicate.

In April 1898 Tyndale White brought the Countess to Williams's London office; she asked him to try to renegotiate the concession with Rhodes. Williams in turn asked her to assist him in any way she might be able to through her various contacts in exchange for a half-interest in any concession he might thereby acquire.[68] After discussing the matter with Rhodes, Williams formally applied to Chartered for a new concession, proposing that a small company capitalized at £100,000 be formed to work it. He expected that the Duke of Abercorn, then Chartered's President, would provide £20,000 working capital, with vendor's shares being divided among the Zambesia Exploring Company, White, and Williams.[69] Chartered agreed to cancel the old concession, replacing it with one of 2,000 acres with all surface rights so located as to include the terminus of the Cape-to-Cairo Railway on the south shore of Lake Tanganyika. They also gave White the right to locate within that area on open ground 1,000 mineral claims in which Chartered retained a 30 per cent interest. In return White had to agree to float a steamer on the Lake and to find £20,000 working capital within six months.[70]

Abercorn did not live up to expectations regarding the working capital, and Williams had to arrange for an extension of time to find

[64] Williams to James Murray, ZEC Director, 10 Oct. 1895, Williams Papers.
[65] Williams to Murray, telegram, 15 Oct. 1895, Williams Papers.
[66] Ibid., telegram, 21 Oct. 1895, Williams Papers.
[67] Dawson, Chapter VII, pp. 7–8.
[68] Ibid., Chapter VIII, pp. 34–45.
[69] Williams to ZEC, 16 May 1898, ZE 4.
[70] BSAC to Williams, 26 May 1898, Williams Papers.

it.[71] Acting on Rhodes's advice, and probably under considerable pressure from him, Chartered agreed to a six-month time extension, but altered the concession slightly. White had to locate his 1,000 claims within two years, but not necessarily within the 2,000 square mile area of exclusive prospecting rights. Chartered's interest in the claims was increased to 35 per cent, and the concession was to be cancelled definitely if the working capital was not found by 26 May 1899.[72] Tanganyika Concessions Limited (Tanks) was formed on 20 January 1899 to take over and exploit the concession.[73]

Even before the time limit for finding the working capital had been extended, Williams began arranging to locate the mineral claims. He asked George Grey (younger brother of Sir Edward Grey, Foreign Secretary, 1905–16, and later Viscount Grey of Fallodon) who had been working for him in Africa to lead the expedition in return for which Grey was to be made African Manager of the exploiting company.[74] After Tanks was formed Grey also received a share in the claims located.[75] Williams also wanted to send Michael Holland, nephew of the former Colonial Secretary, Lord Knutsford, and son of the Prince of Wales's chaplain, who had had some useful experience in Mexico and the Gold Coast, to Lake Tanganyika to act as the company's agent there, and to supervise the reconstruction and launching of the steamship *Cecil Rhodes* which was being built at the Forrest works in Wyvenhoe, Essex. Williams argued that if both men

[71] 'Entry for 21 October in Mr Williams' Diary, 1898', Williams Papers.

[72] BSAC to Williams, 24 Nov. 1898, Williams Papers.

[73] Williams to ZEC, 29 Nov. 1898, ZE 4. The share issue was underwritten by Hirsch and Co.; Baron Hirsch, an Austrian financier, had been befriended by the Prince of Wales. On Hirsch's death in 1896 the firm was headed by Ernest Cassel, close friend of and financial adviser to the Prince. Rhodes and Beit took up some shares but sold them soon afterwards, although the African Trans-Continental Telegraph Co. held some for a longer time. C. J. Leyland who was to become a major figure in the early days of the Benguela Railway took up a large number of shares, while the Belgian Banque Lambert took a small interest, certainly with Leopold's knowledge if not on his behalf. Some funds also came from the Zambesia Exploring Co. in the form of advances to White or to Tanks. Hirsch and Co. completed their underwriting agreement by taking up £3,500 shares in Oct. 1900. (Williams to L. Neumann, private & confidential, 26 Jan. 1899, Williams to Leyland, Private & Confidential, 26 Jan. 1899 and 30 Oct. 1900, TC 25; J. R. Harris to Williams, 17 Feb. 1899, Williams Papers; Williams to ZEC, 2 and 7 Feb., 3 and 10 May, and 16 June 1899, ZE 4 and 1 Nov. 1901, ZE 3; *Morphologie des groupes financiers*, 2nd ed. (Centre de Recherches et d'Information Socio-Politique, Brussels, no date), p. 333.)

[74] Williams to Grey, 23 Oct., 1898 and Williams to ZEC, 18 Nov. 1898, ZE 4.

[75] 'Memorandum of Agreement between TCL and Mr George Grey', copy in the Williams Papers.

left for Africa as soon as possible, information could start flowing back to London quickly. The investing public would then be aware of the possibilities of the concession area before the small firm was transformed into a large company for which Williams had some rather grandiose plans.[76] Although these plans were never realized in their original form, the activities of Grey and Holland in the concession area, and later in Katanga, were to lead to the opening up of some of the world's richest mineral deposits.

[76] Williams to ZEC, 10 Feb. 1899, ZE 4.

CHAPTER II

The Road to Katanga

BECAUSE the Congo–Zambezi watershed seemed geographically similar to the Transvaal goldfields, Williams told Grey to locate Tanks's claims as close as possible to the Congo frontier.[1] He was to be on the lookout for copper, which although less precious and romantic a metal than gold, was increasingly in demand for the expanding electrical industries of Europe and America; the price of copper was £52 a long ton in 1897, £76 two years later. Not surprisingly, then, when Grey reported in September 1899 that, acting on Williams's advice and information from local Africans, he had found a rich copper deposit at Kansanshi, Williams was very enthusiastic. The fact that appreciable amounts of gold were also found in the area made Grey's discovery even more interesting.[2] The gold found was in fact not sufficient to justify full-scale working, although subsequent investigations did confirm the existence of large amounts of copper at Kansanshi. Because of the composition of the ores, however, which present difficult technical problems for treatment, the high expectations for Kansanshi have never been realized. Its lasting importance was as a focal point for initial development and interest in the area, and a stimulus for further prospecting. Much greater ramifications came from Grey's unofficial visits across the border into Katanga.[3]

THE KATANGA CHESS-BOARD

Before issuing Grey's instructions, Williams had discussed Katanga with Albert Ochs,[4] and in June 1899 asked him to help secure a concession to locate between 500 and 1,000 mineral claims there for Tanks, on terms similar to those granted by Chartered.[5] Ochs passed

[1] Williams to Grey, 3 Mar. 1899, ZE 4.

[2] L. H. Gann, *A History of Northern Rhodesia; Early Days to 1953* (Chatto & Windus, London, 1964), p. 121 n.

[3] *Comité Spécial du Katanga, 1900–1950* (CSK, Brussels, 1950), p. 24.

[4] Williams to Ochs, 1 Mar. 1899, Williams Papers.

[5] Ibid., 7 June 1899, Williams Papers. Ochs Bros. (London) to Ochs (Brussels), telegram, 9 June 1899, CCCI Files.

this request on to Thys with a remark to the effect that Tanks's financial arrangements were somewhat laughable—he did not elaborate on this point—but that Williams's group was active and 'not without influence'. In any case, Ochs saw no reason not to accept Williams's money.[6] Thys and Leopold apparently agreed with Ochs, at least on this last point, as later that month Ochs, Williams, and Samson Wiener met to discuss a more detailed proposal. Williams wanted the Congo State to give Tanks a concession of a 2,000-square-mile block of land in Katanga. Tanks would raise £20,000 working capital for prospecting, and register a company in London to locate the block. The Compagnie du Katanga would receive 25 per cent of the vendor's shares in the new company, the right to take up 35 per cent of the issued capital, and could appoint two out of five directors to the company's Board.[7]

No decision on this plan was made, and in January 1900 Ochs urged Thys to 'strike the iron while it [was] hot' and 'not to refuse the enthusiasm of people . . . willing to spend their capital in prospecting.'[8] Ochs and his own group of companies had by this time become more directly involved in the proposed business, despite some continuing reservations Ochs had about Tanks's financial structure.[9] Following an interview with Thys, Ochs proposed a somewhat larger scheme; three 2,000-square-mile blocks on each of which the Ochs/Williams group would have exclusive prospecting rights for ten years, and for each of which they would raise £20,000 working capital.[10] The Compagnie du Katanga would receive fully paid 20 per cent of the issued capital of the British companies formed to exploit any minerals discovered, and would have the right to subscribe 20 per cent of the working capital. Ochs pressed for an immediate decision; if the matter could be arranged by March, an entire year could be gained by taking advantage of the dry season.[11]

Williams took over the leading negotiating role from Ochs, and was soon able to report that the King 'favourably entertained' his ideas, but no decision was forthcoming. In March Williams and Ochs were joined by Thys's son-in-law Gaston Perier, who, in a new

[6] Ochs to Thys, 10 June 1899, CCCI Files.
[7] Williams to TCL, 29 June 1899, Williams Papers.
[8] Ochs to Thys, 15 Jan. 1900, CCCI Files.
[9] Ibid., 27 Jan. 1900, CCCI Files; Agreement between Ochs Bros. and Williams, 31 Jan. 1900, Williams Papers.
[10] Ochs Bros. to Thys, 24 Jan. 1900, CCCI Files.
[11] Williams to Ochs Bros., 16 Feb. 1900, Williams Papers.

arrangement, largely supplanted Ochs both as negotiator and as financial participant.[12] A few days later Williams told Perier that the concession agreement was 'all in order with the exception of some insignificant details' which had to be arranged between the Congo State Government and the Compagnie du Katanga.[13]

Far from 'insignificant', the details in question were in fact of major importance, and a serious obstacle to the conclusion of any agreement for prospecting in Katanga. The basic difficulty was the original provision for the division of the wealth of Katanga between the company and the State. Article 10 of the 1891 convention establishing the Compagnie du Katanga provided for the division of the country into blocks six geographical minutes square. The company was to have the profits drawn from the exploitation of every third square.[14] On a map annexed to the convention, these squares were marked out in a blue and white chess-board pattern. This novel idea was easy to carry out on paper, but had the company ever proceeded to the actual delimitation of the chess-board on the ground, they would have been faced with an enormously expensive and time-consuming task. In any event, the problems involved in trying to divide profits according to the agreement would have been virtually impossible to resolve. After receiving Cornet's report, both the company and the State lost interest in southern Katanga, and in 1896 the company returned to the State all the concession land north of 5° south latitude in return for a concession in the Lomami Valley. Further insistence in 1899 that the company begin the demarcation of the chess-board met with no practical results. The point about the chess-board arrangement that was of most significance for Williams was that there was no provision for granting concessions to third parties.[15]

A situation bearing some similarities to the Chartered–Congo State rivalry of 1890 was beginning to develop. After almost ten years of unproductive activity on the part of the Compagnie du Katanga, Williams—a friend and associate of Rhodes—was prospecting near Katanga, and putting pressure on the company and the State to be allowed to extend his activity into their territory. There had been

[12] 'Agreement . . . Williams, Gaston Perier, and the Suffolk Trust' (Ochs), 10 Mar. 1900, Williams Papers.

[13] Williams to Perier, 21 Mar. 1900, CCCI Files.

[14] H. Waltz, *Das Konzessionwissen in Belgishen Kongo*, 2 vols. (Gustav Fischer, Jena, 1917), i. 122–6.

[15] *Comité Spécial du Katanga*, pp. 21–4.

little enthusiasm among Belgian capitalists for investment in the area, and it appeared that if southern Katanga was to be developed at all, Williams, or some other person who could bring in foreign capital would have to be involved. At the same time it was still essential for Leopold to prevent ultimate control of Katanga from slipping out of his or Belgian hands.[16] Williams agreed to a request not to press his demands while the company and the State resolved their difficulties. This did not prevent his continuing to feed Thys details of what Grey was finding both officially and unofficially on both sides of the frontier.[17] Some of this information inevitably leaked out to the public, the consequent gold fever arousing considerable enthusiasm and comment.[18] Leopold, however, remained most cautious, insisting, without much success that news of Grey's having found gold in Katanga should be kept secret. Sir Francis Plunkett (British Minister in Brussels, 1893–1900) said that Leopold apparently hoped to avoid a 'boom' situation such as had developed following major gold discoveries in the Transvaal in 1886. The King seemed particularly concerned to prevent a rush of British prospectors from the south into Katanga, a rush which the Congo State would not be able to control.[19]

The chess-board problem was solved fairly simply in June 1900 when the Congo State and the Compagnie du Katanga reached an agreement setting up the Comité Spécial du Katanga (CSK). The CSK was, with virtually unlimited administrative powers, to assure and direct the exploitation of all the territory in Katanga south of 5° south latitude and east of a line drawn from the intersection of that parallel with 24°10′ east longitude to 6° south latitude, thence straight south to the State's southern boundary. The State would receive two-thirds of the profit from any exploitation of this area, while the Compagnie du Katanga got one-third. The CSK itself was to be made up of four members, two, including the President, nominated by the State, two by the company.[20]

The CSK did not proceed immediately to consideration of

[16] *Comité Spécial du Katanga*, pp. 21–4.

[17] Williams to Perier, 21 Mar. 1900, Ochs to Perier, 24 Mar. 1900, and Ochs to Thys, 25 May 1900, CCCI Files.

[18] *Comité Spécial du Katanga*, p. 23; Ochs to Perier, 17 and 24 Mar. 1900 and Williams to Thys, 3 Apr. 1900, CCCI Files.

[19] Plunkett to Salisbury, No. 18 Africa, confidential, 11 Jan. 1900, and No. 182 Africa, very confidential, 14 July 1900, FO 123/402.

[20] Convention of 19 June 1900, cited in *Comité Spécial du Katanga*, pp. 26–7.

Williams's schemes. The interest in Katanga which had increased the pressure on Leopold and Thys to solve the chess-board problem also tended to arouse opposition to Williams and the idea of allowing any foreign capital at all into Katanga. One extreme expression of this attitude was voiced by the editor of a minor financial journal, *La Finance Illustreé*, Jules Savygnac, who was also a shareholder in the Compagnie du Katanga. Both by letter and at the company's general meeting called to approve the CSK convention, Savygnac cast aspersions on Williams's character, questioned the existence of gold in Katanga, and generally attacked the company for not having published more detailed information about the possibilities for development and investment in the region.[21] Williams strongly protested against the accusation that he had 'concocted' the story of Katanga minerals simply to create a boom in Katanga shares.[22] A second general meeting—which had to be called as the first had not had a quorum—approved the convention in Savygnac's absence.[23]

After the CSK completed their initial organizational arrangements, they were faced with a vague plan for operations in Katanga put forward by Thys's Banque Coloniale.[24] The bank itself has disappeared leaving little trace, and one can only speculate on what Thys really had in mind and whether he hoped at this stage to be able to keep Williams out of Katanga. Lady Warwick once again entered the picture by introducing Williams to Baron Wettnall, Belgian Minister in London from 1894 to 1903.[25] There is no evidence to indicate the precise nature of the role played in the business by either the Countess or the Minister; no record exists of any direct contact between Lady Warwick and Leopold at this time, but there can be no doubt that her intervention was in some way important. The CSK took up Williams's request, and the Countess received a half-interest in the concession Williams ultimately acquired.[26] He later arranged for Tanks to take up her interest in exchange for 30,000 shares.[27]

[21] Savygnac to Thys, 27 July 1900 and Perier to Williams, 31 July 1900, CCCI Files.

[22] Williams to Katanga Co. [*sic*], 9 July 1900, Williams Papers.

[23] Perier to Williams, 31 July 1900, Williams Papers.

[24] Private information.

[25] Swann, Chapter IX, p. 2.

[26] Williams to ZEC, 14 Feb. 1901, Williams Papers.

[27] Williams to Leyland, 12 Sept. 1901, TC 25; Williams to ZEC, 14 Feb. 1901, Williams Papers.

The CSK and Williams reached an agreement, signed on 8 December 1900, for the joint mineral exploration of Katanga for a period of five years, with Williams providing up to £5,000 annually for the purpose.[28] Williams's prospecting right was extended by an agreement of 22 October 1905 for a further four years. One writer maintains that it was relatively easy for Williams to get this concession because of the prevailing pessimism in Belgian circles about Katanga's mineral potential.[29] How easy Williams's task had in fact been may be questioned, but the pessimism is reflected in the provision that the CSK would only spend up to £3,000 annually after Williams's contribution had been used, and if such additional expenditure was deemed essential. If Williams wanted to spend his money the Belgians had no objection, but would only spend their own when they were convinced of the value of the mineral deposits. Any companies formed to work deposits were to be capitalized equally by Williams and the CSK, with Williams getting 40 per cent of the profits, the CSK 60 per cent. Half the companies formed were to be registered in Brussels, half in London. Joint exploitation was to continue for 30 years, subsequently extended to 89 years on condition that after 30 years all the exploiting companies would increase their capital by 30 per cent, turning over the entire amount of the increase to the CSK.[30] Williams could transfer his interest to a company on condition that the company's statutes were approved by the CSK who would have the right to appoint two of the company's directors.[31] In December 1901 Williams transferred his concession to Tanks.[32]

Immediately after signing the Agreement, Williams cabled Grey to organize a trip into Katanga.[33] Plans had already been made for a second expedition to Northern Rhodesia to go more thoroughly over the area already investigated, but had been delayed by Lord Kitchener's closure of the railways in South Africa during the Boer

[28] Agreement between Williams and Hubert Droogmans as President of the CSK, 8 Dec. 1900, copies in Williams Papers. An account, with some minor inaccuracies, of this agreement and its subsequent modifications can be found in: Keith, pp. 90–4. See also J. A. Bancroft, *Mining in Northern Rhodesia* (British South Africa Co., London, 1961), pp. 50–1.

[29] H. C. Gunning, 'The Early Years' in F. Mendelsohn (ed.), *The Geology of the Northern Rhodesian Copperbelt* (MacDonald, London, 1961), p. 7.

[30] Droogmans to Williams, 14 June 1901, Williams Papers. After 1905, expenses were shared equally.

[31] Agreement . . . 8 Dec. 1900.

[32] Droogmans to Williams, 3 Dec. 1900, Williams Papers.

[33] Williams to Droogmans, 12 Dec. 1900, Williams Papers.

War.[34] Grey's second expedition was therefore only able to start northwards from Bulawayo in April 1901, first spending several months in Northern Rhodesia before crossing into Katanga.[35] Another party under Michael Holland left from Abercorn in May, by which time Holland had supervised the reconstruction and floating of the steamer *Cecil Rhodes* on Lake Tanganyika as required by the terms of the concession.[36] This task had not occupied all his time, leaving him free to participate in several punitive expeditions with Robert Codrington, then Administrator of Northeastern Rhodesia, against Africans who were still selling slaves. One of these expeditions, in October 1899, was against Chief Kazembe, in the region around Lake Mweru, whose rumoured ivory cache Holland hoped to secure for Tanks. He failed to get the ivory, but did gain some useful information which subsequently helped him locate copper mines in Katanga. From June to August 1901 Holland travelled over the Katanga concession area examining deposits and mines, most of which had already been noted by Cornet.[37]

In August Holland and Grey joined forces. Grey was convinced that Holland's discoveries clearly indicated the existence of valuable copper deposits, a view supported by the expedition's mineralogist, Franz Studt, and later by J. R. Farrell, an American mining engineer who visited the mines in 1902 and reported to Williams on their mineral content and commercial possibilities.[38] Before the mines could be worked successfully, a number of practical problems had to be overcome. One discussed by Grey was finding a sufficient number of workers for the mines and treatment plant. Of more immediate importance, and requiring considerably more finance, was the need for cheap transport facilities. While Cornet's appreciation of the value of the deposits may not have been entirely accurate, there was no doubt that the distance between the mines and the sea created considerable problems for bringing in materials and supplies and

[34] Williams to ZEC, 20 Mar. 1901, ZE 3.

[35] Grey to Williams, 3 Jan. 1903, in 'Report on the Discoveries made by . . . Grey's Expedition . . .' printed by Tanks for circulation to shareholders.

[36] The SS *Cecil Rhodes* only saw a few years' service before running aground on a sandbank. She was not refloated and was finally cut up and removed by German forces during the First World War.

[37] Memorandum by Holland written in 1925, Williams Papers; C. G. Watson, one of Williams's representatives in Katanga, to Williams, No. 173, 16 June 1917, TC 138A.

[38] Grey to Williams, 6 Sept. 1901, and Farrell to Williams, 6 and 8 Jan. 1901, in 'Report . . .', op. cit., pp. 25–7, 89–129.

shipping out copper and ores. If the mines were to be commercially viable, a railway had to be built linking them to the coast.

THE SOUTHERN ROUTE IS CLOSED

As the same was true of Kansanshi, whose essential worthlessness had yet to be appreciated, Williams had already talked to Rhodes about the possibility of altering the proposed route of the Cape-to-Cairo Railway to help open up 'one of the most valuable permanent mines in Rhodesia'.[39] In April 1900 Rhodes suggested that, as Williams seemed to be getting on so well with Leopold, it might be possible for him to arrange for the Cape-to-Cairo Railway to go through the Congo. Williams agreed to negotiate this with the King in return for the right to link his mineral area to the main line of the railway in Rhodesia. He also wanted to be able to locate the township originally planned for the southern shore of Lake Tanganyika near Kansanshi if that seemed preferable.[40] Chartered saw no objection to this, and gave Tanks the right to build the link 'upon terms to be agreed upon',[41] and modified the original concession to allow Tanks to chose where the township would be located.[42]

Before negotiations between Leopold and Chartered had been broken off in 1891, the King had told Thys to raise the question of connecting a railway line from Katanga to the railway that Chartered was required to build to its northern frontier.[43] The value of such a connection was apparent; the difficulty was whether or not Rhodes and Leopold could agree on terms. Williams discussed the railway situation with Hubert Droogmans who in addition to being President of the CSK was very close to the King. Williams emphasized the advantages for Katanga of a line to the south; the cost of expeditions to Katanga would be cut tremendously, great resources could be opened up, and the last remnants of the slave-trade wiped out. More specifically Williams wanted a concession to build a railway through the Congo to the Nile. In exchange for various mineral rights along the route he was prepared to connect the southern boundary of Katanga to the Mashonaland Railway in Rhodesia within five years. He would then carry on construction through Katanga at the rate of

[39] Williams to Rhodes, 24 Apr. 1900, Williams Papers.
[40] Dawson, Chapter X, pp. 6–7.
[41] Rhodesian Railways, Ltd., to Williams, 15 May 1900, Williams Papers.
[42] BSAC to TCL, 15 May 1900, Williams Papers.
[43] Leopold to Thys, 3 Jan. 1891, Thys Papers, Vol. xxviii, No. 1.

100 miles per year, at a cost of about £6,000 per mile.[44] Such a scheme could serve Katanga, Kansanshi, and Leopold's desire to use the navigable portions of the Congo River and give the north-eastern Congo an outlet on the Nile.

On 12 June 1901 the first step towards building this line was taken when Leopold and Williams signed an agreement for the formation of a Congolese company to survey a railway route from a point on the southern boundary of the Congo to a point on Lake Albert, at or near Mahagi. A separate construction company was to be formed after the survey was completed. A through route to the Nile was not yet being considered specifically, but this proposed line was intended to use some sections of the navigable Congo River that might form part of such a route. The Congo State could determine the point at which construction would begin, and construction was to proceed in 300-kilometre sections, no new section being begun until the previous section had been opened to the public.[45] This last provision would inevitably slow down construction, but was probably a manifestation of financial caution. Williams offered Lady Warwick an interest in the construction company, but it is not clear whether or not she accepted.[46]

The necessary co-operation between Rhodes (or Chartered) and Leopold was not easy to arrange. Christopher J. Leyland, a major shareholder in Tanks and the Zambesia Exploring Company, and general financial backer of Williams's enterprises, was afraid of the effect Rhodes's involvement might have on negotiations.[47] Lady Warwick was more blunt, if somewhat romantic; in her view, having dealings with Rhodes 'was as much as Leopold's throne was worth'. Williams was sure that Rhodes and Beit would ask him to get enormous concessions from the Congo in return for their agreeing to extend Chartered's railway to the Katanga frontier. He also thought it likely that Leopold would not give the concessions.[48] He was right in both cases.

Rhodes and Beit, still wanting to control Katanga's minerals, wanted both Williams and Leopold to turn over half their Katanga interest to Chartered. Tanks was also to raise its capital from £130,000 to £430,000, giving Chartered the entire amount of the increase. Beit

[44] Williams to Droogmans, 3 May 1901, Williams Papers.
[45] Agreement between the Congo State and Williams, 12 June 1901, Williams Papers.
[46] Williams to [Lady Warwick], 12 July 1901, Williams Papers.
[47] Leyland to Williams, 11 Sept. 1901, TC 25.
[48] Williams to Leyland, 4 Oct. 1901, TC 25.

and Rhodes would then buy £150,000 of Lady Warwick's shares.[49] In discussions with Leopold, Williams supported his claim that the rail link to the south was of paramount importance by telling the King of a telegram he had just received from Grey reporting enthusiastically on Holland's discoveries. This backfired somewhat as it apparently confirmed the King in his intention to connect the mines to the navigable Congo, despite Williams's insistence that copper transported over this route could not be made to pay.[50] Williams was somewhat depressed by this reaction, but saw the future in glowing terms; this was, he felt, an important turning-point in his negotiations.[51]

The real turning-point came later. Leopold and many of the men around him were still not entirely convinced of the mines' value. Baron van Eetvelde went so far as to say that he did not believe Grey's and Holland's reports; the Congo would stick to its rubber. Williams retorted rather angrily, and stormed out of van Eetvelde's office. The strength of his reaction seemed to have had some effect on Leopold, and shortly afterwards, Henri Buttgenbach, a Belgian engineer, was to investigate. It was only after his reports had been made and published in June 1902 that the last remnants of scepticism disappeared.[52]

Acceptance of the commercial viability of the mines made the railway question more urgent. As expected, Leopold refused to accede to Rhodes's demands. Discussion of a link to the South was dropped,[53] while arrangements for the construction of a line to the navigable Congo and Mahagi continued. On 11 March 1902 the Compagnie du Chemin de fer du Katanga (CFK) was formed, initially to survey a possible route to put into effect the June 1901 Agreement which was modified to allow the same company to construct and operate the railway after the survey was completed. It was only in March 1903, when Buttgenbach had made his report on Katanga, that an expedition was organized, under the leadership of Commandant Jacques, to start the survey.[54] Also in 1902 the

[49] Beit to Williams, 17 May 1902, TC 25.

[50] Dawson, op. cit., Chapter XI, pp. 10–12.

[51] Williams to his wife, 17 June 1901, extract in Williams Papers.

[52] Dawson, Chapter XI, pp. 21–2; *Mouvement Géographique*, 30 Nov. 1902, cols. 577–88; G.-P. Perier, *Les Flèches du Congophile* (Éditions l'Afrique et le Monde, Brussels, 1957), pp. 65–6.

[53] Williams to Rhodes, telegram, 19 and 20 Oct. 1901, TC 25.

[54] P. Sorel, 'Les Origines et les débuts de la Compagnie du chemin de fer du Katanga', in *Comptes Rendus du Congrès Scientifique* . . ., vii (CSK, Brussels, 1950), pp. 34–5.

Compagnie du Chemin de fer des Grands Lacs Africains (CFL) was formed. This company built the rail lines linking the navigable portions of the Congo River between Bukama and Stanleyville.

By the time Buttgenbach reported, circumstances had altered considerably. Rhodes had died and Williams's relations with Chartered very much deteriorated. There would appear to have always been some animosity between Beit and Williams, possibly because Beit had also wanted to assist Lady Warwick and secure her support for some of his own activities, but there were some specific reasons for this deterioration as well. One was an interview Williams gave to a Reuter correspondent with, he maintained, Chartered's full accord.[55] Under the headlines 'Cape to Cairo—German Route Abandoned—Agreement with Belgium', the *Daily Telegraph* printed an article based on this interview, outlining the June 1901 agreement, and pointing out that this meant the Cape-to-Cairo Railway would run through Congolese territory rather than through German East Africa, as it was generally expected to do. Chartered, denying any prior knowledge of what Williams had intended to say, protested against this conclusion rather hotly.[56] In the ensuing dispute, Williams said that he thought he could strengthen his own position by publishing the entire history of his previous negotiations with Rhodes. These, he maintained, would show that Rhodes had intended the line to go through the Congo in the first place, and that it was Beit who had brought negotiations to an end by insisting on a half-interest in all Katanga minerals. At the same time, Williams said that his feelings of loyalty to Rhodes put him in a 'peculiar' position as such publication would show Rhodes to have been 'double-dealing'.[57] He said nothing of any feeling of loyalty or discretion regarding Lady Warwick.

While Williams may very well have been sincere in his feelings of loyalty to Rhodes, there is no reason to think that publication of these details would have been of any advantage to him at all. He agreed to publish nothing further on the point, Beit and Jameson agreeing not to publish any more about the Cape-to-Cairo line going through German East Africa. Williams was caught between Chartered who wanted a large share of Katanga's wealth, and Leopold who

[55] Memorandum '. . . Williams' negotiations re . . . Railways', undated, Williams Papers.

[56] J. Jones, Secretary, BSAC to Williams, 29 Apr. 1902, Williams Papers.

[57] Williams to Droogmans, private, 26 May 1902, TC 25.

wanted to keep that wealth under his own control and furthermore had plans for using the Congo River to that end although Williams was convinced this could not be done profitably. Unless the conflict between Leopold and Chartered could be resolved, Williams was afraid that both Kansanshi and his Katanga interests would prove worthless. Despite his conviction that Chartered would not remain so intransigent, he decided to seek an alternative transport route.[58]

THE PORTUGUESE CO-OPERATE

The first step in this search was to send an expedition to Angola under the leadership of Cecil Learmonth, who was originally told to investigate possible overland routes for sending materials to Kansanshi and Katanga for further prospecting and development work.[59] The underlying intentions are clearly indicated in a scheme Williams outlined for building a railway between Loanda and Katanga. The Congo State refused an invitation to participate in the project, as they felt they had first to complete what they had already undertaken.[60] Williams shifted the focus of his scheme from Loanda to Lobito Bay and Benguela, most probably because Lobito Bay was an excellent natural harbour, and the Portuguese had for several years been intending to build a railway from Benguela to the Angolan highland plateau, but had had no success in finding the necessary funds. Learmonth returned from Africa in December 1902 with the suggestion that it would be better to build the line from Mossamedes, further south, because of the very difficult terrain between Lobito and Caconda.[61] The existence of prior concessions to German and Anglo–German companies operating in southern Angola made this impossible.[62] By the time Learmonth made the suggestion, the time when Williams might have tried to take it up had in any case passed.

In London Lady Warwick was still at work earning her half-interest in Williams's schemes. She arranged for Williams to be introduced to the Portuguese Minister in London, the Marquês de Soveral, well known and liked in London society and a personal friend of King Edward. The introduction was made by Mr. Henry (later Viscount) Chaplin, MP for Lincolnshire and Wimbledon

[58] Williams to Droogmans, 4 and 12 June 1902 and T. Heyvaert, CFK President, to CSK, 7 June 1902, TC 25.

[59] Ibid.

[60] Droogmans to Williams, 9 June 1902, TC 25.

[61] Gosselin to Lansdowne, No. 84 Africa, 24 Dec. 1902, FO 179/368.

[62] See below, p. 42.

constituencies between 1868 and 1916; his brother-in-law was the 4th Duke of Sutherland who was married to Lady Warwick's half-sister. After meeting Soveral a few times Williams wrote on 16 June 1902:

> I am desirous of securing a concession from the Portuguese Government for the right to build a railway either from St. Paul de Loanda or Benguela towards the eastern boundary of Angola, and to have mineral rights over the territory through which the railway would pass. In other words, I should like to have a concession to build all railways and have the sole right to minerals within 8° and 14° south from the coast to your eastern boundary.
>
> I shall be glad if you can assist me to acquire this, as I feel certain that, if such a concession is granted to me on terms satisfactory to your Government and myself, the result would be mutually beneficial.[63]

While preliminary action was being taken on this request, Williams tried a new approach to the problem of linking the Katanga mines to the Rhodesian rail system. The Congo State remained unwilling to grant Chartered any concession, but showed some willingness to participate in the financing of Chartered's line to the frontier.[64] Williams then offered to build a railway from the point to which Chartered was at the time planning to build their line—Broken Hill—to the Katanga border. This was on condition that an amicable arrangement was reached by all concerned, and that he would have the right to connect the line to the CFK.[65] His extreme optimism that Chartered would accept the idea[66] proved completely unjustified. It was only after Williams was in financial difficulty over the Benguela Railway and Chartered was having problems supporting the line between Victoria Falls and Broken Hill, that Chartered showed any interest whatever in reaching an understanding with him and the Congo State.[67]

On 15 October 1902 Williams arrived in Lisbon, where Soveral introduced him to the Acting Minister of Foreign Affairs, Fernando

[63] Dawson, Vol. ii, Chapter I, pp. 9–10; Tanganyika Concessions Limited, annual report, 1905, p. 7; José d'Almada, *Para a Historia do Caminho de Ferro de Benquela*, (BRC, Lisbon, 1951), pp. 35–6; P. Magnus, *King Edward the Seventh* (John Murray, London, 1964), p. 285. Williams's letter to Soveral is quoted in d'Almada, p. 36.

[64] Droogmans to Williams, 23 June 1902, TC 25.

[65] Williams to RRL, 11 July 1902, Williams Papers.

[66] Williams to Droogmans, 11 July 1902, Williams Papers.

[67] See below, pp. 66–74.

Mattoso Santos, who in turn introduced Williams to the Overseas (i.e. Colonial) Minister, Antonio Teixeira da Souza. After some discussion Teixeira de Souza told Williams to put his scheme in the form of a petition; for advice on this the Minister sent Williams to a lawyer, Senhor Alves de Sà. As Williams and de Sà had no common language, a British subject, D. M. Lane, who was a shipping agent in Lisbon, was called in to act as interpreter. De Sà did not think the concession could be arranged in less than a year, and that even then it would be the fastest thing of its kind ever done in Portugal.[68] Williams persisted, and by 22 October was able to tell Leyland that the railway concession would be granted with the right to exploit minerals in strips of land 100 miles wide on both sides of the line.[69]

Negotiations had been carried out in considerable secrecy at Teixeira da Souza's home, although rumours of what was being discussed circulated in Lisbon for many weeks.[70] Early in November one newspaper reported that 'money was already running into certain pockets'. It is not unlikely that some 'added inducement' was in fact offered to ministers and others involved to put through such an important agreement in what would be considered anywhere a short time, and Williams was probably not averse to 'squaring' people if the need arose. Not surprisingly, there is no firm evidence supporting the newspaper's claim, although Lane did refer in a letter to a draft drawn by Williams that was cashed by a London banking house so that it would not pass through any other bank or firm in Lisbon.[71] Only de Sà, Lane, and Williams were supposed to know about the transaction. The ministerially inspired *Novidades* acknowledged the rumours by referring to an 'embroyo and necessarily reserved agreement'.[72] In addition to any financial and social influence involved, there was also royal interest in Williams's scheme. In late November Don Carlos, King of Portugal, was visiting King Edward VII at Windsor. The two monarchs discussed the proposed concession and used their influence to ensure that it was granted. On 28 November 1902 Queen Amélie, Regent in her husband's absence, having received telegraphed instructions from the King, signed an agreement giving Williams the right to build a railway from Lobito

[68] Dawson, Vol. ii, Chapter 1, p. 19.
[69] Williams to Leyland, 22 Oct. 1902, TC 25.
[70] Gosselin to Lansdowne, 2 Dec. 1902, FO 800/139.
[71] Lane to Williams, 6 Nov. 1902, enclosing translation of article headlines 'It is True' from *Fotha da Jarde*, 4 Nov. 1902. TC 25.
[72] 8 Nov. 1902; translation in Lane to Williams, 10 Nov. 1902, TC 25.

Bay to the eastern frontier of Angola via Benguela.[73] Immediately after signing the agreement, Williams telegraphed the news to King Edward.[74] He is reported to have told a group of young officers from a British ship then visiting Lisbon, with more accuracy than discretion, 'I have floated Lady Warwick',[75] to whom he later gave the pen he had used to sign the concession.

The agreement provided that within six months of signature Williams had to present for government approval the articles of association of a Portuguese limited liability company which would have the right to build and operate a railway from Lobito to the Congo border for a period of 99 years. In addition to mineral rights in the territory 120 kilometres on either side of the railway line, the company was also to have a variety of commercial, industrial, and tax privileges. Portuguese concern for keeping control of its colonial resources firmly within its own hands was clearly shown by the requirement that the company have its head offices in Lisbon and that the majority of its directors be Portuguese nationals. The company had to build a telegraph line along the railway, the facilities of which would be available to the government for official use free of charge. Williams had to put up a deposit of £100,000 as a guarantee that the time provisions of the agreement would be met; if they were not, except in case of *force majeure*, the deposit would be forfeit. The first 400 kilometres of line had to be opened to traffic by the end of 1905. After ten years the government could, at six months notice, redeem the railway, but not the 120-kilometre strips of land, the redemption price to be determined by taking the company's net proceeds during the seven years prior to redemption, subtracting the proceeds of the two least productive of those years, and taking the average of the remainder as an annuity to be paid to the company during the remaining years of the concession.

The agreement met with generally favourable reactions in Portuguese commercial circles. The press, of whatever political tendency, sometimes calling Williams 'O Millionario', came to a generally approving conclusion, while news of the concession was greeted enthusiastically by Portuguese interested in African trade and in Angola.[76] Sir Martin Gosselin, who had only recently arrived in

[73] Alexander to Williams, 13 Sept. 1933, J. J. Machado to Williams, private and confidential, 3 Nov. 1904, Williams Papers; d'Almada, pp. 38–9.

[74] Gosselin to Lansdowne, 3 Dec. 1902, FO 800/139.

[75] Gosselin to Sir F. Bertie, private, 29 Dec. 1902, FO 800/160.

[76] Gosselin to Lansdowne, 3 Dec. 1902, FO 800/139.

Lisbon to serve as British Minister there, was somewhat annoyed at having been told nothing about Williams by the Foreign Office.[77] Although the two men had never met, Gosselin judged from what he heard that Williams 'must be a bit of a bounder'.[78]

In fact the first knowledge the Foreign Office had of Williams's negotiations or of his intentions came to them in Gosselin's telegram informing them of the publication of the concession decree in the *Diario do Governo*.[79] In the circumstances, the view reported by the Congo State's Consul-General in Lisbon that Williams was acting as the agent of the British government which had put up the £100,000 is nothing less than laughable.[80] With a dispatch giving a somewhat fuller account of the concession Gosselin enclosed a note from Williams expressing regret that he had been unable to call at the legation during his stay in Lisbon; Williams said he had had to leave for Paris immediately.[81] Gosselin had been told, however, that the real reason for Williams's failure to contact the legation was that he was afraid of the possible results in the light of the Anglo–German 'Secret' Treaty of 1898.[82]

Soveral had in fact warned Williams not to let anyone in the British Foreign Office know why he was in Lisbon for precisely this reason.[83] The Agreement of 1898 provided that if either Britain or Germany was prepared to offer Portugal a loan secured on the revenues of Angola, Moçambique, or Timor, the other would be informed and given the opportunity of participating. A 'secret' convention attached to the agreement added that in the event of Portugal's deciding to give up these colonies Great Britain and Germany would divide them. Great Britain secured the reversion of southern Moçambique, including Delagoa Bay, and a strip of land in central Angola, including Loanda, while Germany secured southern Angola adjacent to South West Africa and the Otavi tin-mining area, the northern quarter of Angola, and northern Moçambique.[84] This was probably

[77] Gosselin to Bertie, 17 Dec. 1902, FO 800/17.

[78] Ibid., 29 Dec. 1902, FO 800/160.

[79] Gosselin to FO, telegram, No. 29, 29 Nov. 1902, FO 2/638.

[80] Empis, EIC Consul-General in Lisbon, to van Eetvelde, 22 Nov. 1902, EIC, AE 158 (224).

[81] Gosselin to Lansdowne, No. 69 Africa, confidential, 2 Dec. 1902, FO 179/368.

[82] Gosselin to Lansdowne, 3 Dec. 1902, FO 800/139.

[83] d'Almada, p. 39.

[84] For a discussion of events leading up to the conclusion of the agreement, the strategic considerations behind it, and some of its repercussions, see W. L. Langer,

the most widely known secret in diplomatic history. Soveral knew of it before the agreement was signed and urged Salisbury not to sign, while an anonymous columnist in the *Fortnightly Review*[85] discussed it in detail. The agreement was renegotiated in 1913[86] but was cancelled by the outcome of the First World War. Italy proposed that she be allowed to accede to Germany's reversionary rights, but the idea was given short shrift by the Foreign Office.[87]

A large part of the proposed Benguela Railway line ran through what Germany already tended to regard as her territory. Germany also had considerable commercial interest in southern Angola, which would in no way appreciate the competition the Benguela Railway could be expected to produce. It was not at all surprising that Gosselin was particularly concerned about what he should say to the German Minister in Lisbon, Count Tattenbach, if the question were to come up between them. He was instructed to confine himself to the strict truth—that he had had no communication with Williams.[88]

The nature of German opposition based on its commercial interests was complicated by the fact that these were closely linked to Chartered, both in South West Africa and in southern Angola. For years these combined interests had been trying to get the Portuguese to agree to allow them to build a railway from Mossamedes to Porto Alexandre and the Otavi mines. In 1894 an Anglo–German Company, largely financed by Chartered, the Trans-African Railway Company, Ltd., was given a concession to build such a line, but no progress had been made because the government and the company could not agree on the route to be taken. Tattenbach had always taken a strong position in defending German aspirations in Angola and once again sprang to the charge. He maintained that the Portuguese, having given the concession to Williams, should be willing to reach an

---

The Diplomacy of Imperialism, 1890–1902, 2nd ed. (Knopf, New York, 1951), pp. 523–9. See also P. R. Warhurst, *Anglo–Portuguese Relations in South-Central Africa, 1890–1900* (Royal Commonwealth Society, London, 1962), pp. 129–46, and R. J. Hammond, *Portugal and Africa: A Study in Uneconomic Imperialism* (Stanford University Press, Stanford, 1966), pp. 245–71. Many of the relevant Foreign Office documents were reproduced in the FO Confidential Print No. 7303, 'Correspondence respecting Portuguese Possessions in Africa, 1898, Part 1'.

[85] Oct. 1898, pp. 627–34. The agreement was in fact signed by Balfour while Salisbury was out of the country.

[86] See below, pp. 75–6, 80, 105.

[87] The relevant correspondence on this can be found in FO 371/4401/703.

[88] FO to Gosselin, No. 69 Africa, confidential, 2 Dec. 1902, FO 179/368.

agreement with German interests to build a line from Tiger Bay to Otavi as a form of compensation.[89]

Chartered also came to oppose the Benguela Railway as a threat to the Rhodesian railways, while Leopold added to Williams's difficulties by refusing to co-operate by giving assurances that the line would be connected to the Katanga mining area. The obstacles that were to arise were great, but in November 1902 the problem of finding the £100,000 deposit, forming the company, and financing construction presented sufficient difficulties for the time being.

[89] H. Drechsler, 'L'Allemagne et l'Angola du sud, 1898–1903', *Présence Africaine*, xlii (1962), 54; Gosselin to Lansdowne, No. 81 Africa, confidential, 23 Dec. 1902, FO 179/368, and No. 88, Africa, confidential, 1 May 1903, FO 2/769.

CHAPTER III

A Slow Start

THE LADO IMPASSE

Tattenbach was not alone in questioning 'O Millionario's' ability to command sufficient financial resources to bring his railway plans to a successful conclusion. Whether or not he could depended very much on King Leopold; without a firm assurance that the Benguela Railway would be linked to the Katanga mines, Williams's scheme was pure speculation with no attraction for any but the most incautious investor. Leopold refused to make any firm decisions about the Benguela line until the 'Nile question' was settled.[1] Having a rather distorted view of the extent to which businessmen could influence the British government, the King wanted to use Williams, as he had earlier tried to use Rhodes, to help him finally achieve his ambition of securing access to the navigable Nile for the Congo State. He wanted his lease on the Lado Enclave to be perpetual, and his claim to the Bahr-el-Gazal, which French opposition had forced him to abandon in 1894, revived. The British government refused Leopold's request to put the matter to arbitration, but not wishing to appear 'obstructive' or 'dilatory' offered evidence of 'good faith' by agreeing to negotiate.

As the French threat to the southern Sudan and the Nile had been eliminated in 1898 following Marchand's retreat from Fashoda, there was no longer any strategic reason for allowing Leopold into the area. It was thought for a time, however, that some degree of co-operation with the Congo State might be useful in opening up Uganda to trade and administration. It was this view that led to the conclusion in May 1906 of a treaty which gave Leopold the right to build a railway from the Congo frontier to the Nile. The lease on the Bahr-el-Gazal was abrogated and the terms of Leopold's tenure of Lado remained unchanged. A lease on a strip of land between the Sudan/Congo border and the port of Mahagi on Lake Albert was to remain in force as long as the Congo remained under the control of

[1] Droogmans to Robert Williams & Co., telegram, 28 Jan. 1903.

Leopold and his successors. This was all the King could salvage. Belgian business interest in the railway quickly waned, and a survey carried out in 1907 showed that British aims could best be served by building a line through Uganda itself.[2] Not surprisingly he was very dissatisfied with the agreement and his relations with van Eetvelde who had negotiated it became rather strained.

Williams supported Leopold in any way he could, writing letters to the Foreign Office and to editors of magazines and newspapers defending the Congo State against growing international opposition, and trying to secure the support of Lord Cromer. As Williams's experience in connection with the Congo was limited to Katanga which did not come under the fire of those agitating for changes in the State's administration, his efforts on the first point were of little value. On the second they were simply unsuccessful. Williams also approached Chamberlain,[3] who he thought, for some incomprehensible reason, would support him not only in convincing Chartered to be more amenable but also in arranging for a branch line of the Benguela Railway to be built to connect directly with the Rhodesian rail system. Here, as in many other cases, Williams showed himself unable or unwilling to appreciate the political factors involved in his schemes. Although the proposed branch line might have been expected to provide cheaper transport for Rhodesian imports and possibly for some of the coal exported from Wankie than Chartered's own lines, Chamberlain could hardly have been expected to support a railway that operated in Portuguese territory but was opposed by British financial interests already established in southern Angola, and would operate to the detriment of railways in British territory.

Without placing too much emphasis on the branch line, Williams outlined several alternative schemes for finding the £3,000,000 he estimated would be needed to build the approximately 800-mile-long main line. One common feature was the formation of a new company to take over the railway concession and all or part of Williams's share of Katanga minerals.[4] Leopold refused to allow a transfer of Williams's holdings but at the same time apparently said something,

[2] Van Zuylen, *L'Échiquier . . .*, pp. 281–300; Memorandum by A. F. Orchard, Staff Officer, FO Librarian's Department, 'Congo–Nile (Lado) Railway', 12 Oct. 1927, in FO 371/1262/8172.

[3] Williams to Mrs. Williams, 5 Apr. 1903, Williams Papers, and Williams to Chamberlain, 5 Apr. 1903, CO 417/379.

[4] Williams to Droogmans, private and confidential, 4 Nov. 1902, Williams Papers.

or allowed something to be said for him, which made Williams and Leyland believe that his support for the Benguela line would eventually be forthcoming, provided that this was not made public knowledge.[5] There is no direct evidence that such assurances were ever given, but unless they were it is impossible to understand what justification there was for going ahead with attempts to form the Central African Corporation with a share capital of £3,000,000 all of which was to be given fully paid to Tanks. Williams himself was to receive £250,000 in cash to repay £150,000[6] that had been advanced by the Zambesia Exploring Company and the £100,000 Leyland had put up to be used as the guarantee deposit with the Portuguese government. In return the Central African Corporation was to receive the Benguela Railway concession, 1,600 shares in the CFK and all of Williams's interest in the Katanga mines. Leyland agreed to underwrite the issue of £2,500,000 debentures, out of £4,000,000 to be authorized, for a 5 per cent commission. Williams and Leyland were to be life directors of the Corporation, and were to be paid one-fourth of its profits after 35 per cent had been paid on the shares, less one-fourth that was to be applied to debenture redemption. The life directors were also to have the right of veto over any proposed sale of the enterprise.[7]

Even leaving aside the question of whether or not Williams could rely on Leopold's support, it is difficult to understand why he wanted to start yet another company rather than amalgamate all his interests in either Tanks or Zambesia Exploring. If the idea was to ensure the independence of Kansanshi, it would have been much less complicated to float a separate company for it, and certainly much less costly. One possible explanation is that Williams wanted somehow to involve Belgian interests inextricably in the Benguela Railway, thereby protecting it against possible opposition in the future. This did not require the complex arrangements proposed, and one is forced to conclude that personal gain from the formation of the new company played at least some part in the decision.

[5] L. Scotland (Tanks's secretary) to Leyland, 11 Feb. 1903, and Leyland to Williams, 29 Mar. 1903, TC 25.

[6] The precise nature of the £150,000 expenses incurred in connection with the Benguela concession is not given in any of Williams's memoranda or correspondence. Part of the money may very well have been used in connection with the mysterious draft mentioned by Lane (see above, p. 39).

[7] TCL circular to Shareholders, 27 Apr. 1903, and Williams to TCL, 21 Apr. 1902, TC 25; Draft Prospectus of the Central African Corporation, Williams Papers; *The Economist*, 2 May 1903, pp. 779–80.

WORK FINALLY BEGINS

While Williams was in this very weak financial position, Chartered approached him to see if something in their mutual interests could be arranged. One suggestion was that the concession held by the Trans-African Railway Company to build a railway in German South West Africa and southern Angola be amalgamated with the Benguela concession.[8] Chamberlain advised Williams against this as he did not think Chartered could offer advantages equivalent to those Williams could offer them.[9] Why Chamberlain intervened even to this small extent is not at all clear, particularly as Chartered hoped to be able to ensure that the Benguela Railway would never be built, and could more easily have done this if their Angola interests and Williams's had been closely connected.

Williams suggested that Chartered give him the right to build the railway from Rhodesia to Katanga, fair transport rates for copper and coke, and generally support his negotiations with the Belgians. In return he would give them the Benguela Railway provided that they paid all expenses incurred up to that time and carried out the terms of the concession agreement. Alternatively, Chartered could turn over their railway to Salisbury to the Central African Corporation which would then issue debentures to extend the line to Katanga and connect with the Benguela Railway. He thought the first of these suggestions the better in view of the possibility that the Rhodesian railways would some day amalgamate with the South African rail system. Williams went on to say that although he could not give Chartered a direct interest in Katanga, either of these proposals would have brought his and Chartered's interest in developing those minerals and in providing railway traffic together and would not have involved Leopold directly. He pressed for a rapid decision as he expected to proceed forthwith to form the new company. Chartered asked Williams to make a more specific proposal, though it is not clear whether they meant this seriously or merely wanted to temporize. Williams intended to discuss the matter with Chamberlain at the end of April 1903 when Leopold summoned him urgently to Belgium.[10]

Whatever financial arrangements were ultimately to be made, the

[8] Williams to Chamberlain, 1 May 1903, CO 417/379.
[9] Chamberlain to Williams, 1 May 1903, CO 417/379; d'Almada, pp. 42–3.
[10] Williams to Chamberlain, 21, 23, and 26 Apr. 1903, CO 417/379.

terms of the concession agreement required the formation of the Portuguese company to build and operate the railway. On 6 May 1903 a preliminary meeting of what was to be the London Committee of the Companhia do Caminho de Ferro de Benguela (Benguela Railway Company) set the company's share capital at £3,000,000. The directors and auditors divided the first shares issued, with Williams and Leyland getting the largest portion and the Portuguese government being given 10 per cent of them fully paid.[11] Eight days later the proposed statutes of the company were submitted to the Portuguese government for approval, which was by no means certain. There was strong opposition to the railway company but it had some fairly influential support as well. Joachim J. Machado, former governor of Moçambique, whose opinion carried considerable weight in colonial circles, was in favour of it, as was Eduardo F. Pinto Basto, President of the Lisbon Chamber of Commerce.[12] On 25 May 1903 the statutes were approved[13] and the company officially formed the following day. At the first meeting of the Board of Directors, Pinto Basto was elected President, Machado named Managing Director, and Lane appointed Secretary.[14]

Williams returned to his negotiations with Chartered, asking Rhodesia Railways Ltd. (formed in 1899 to build the railway from Bulawayo to the Wankie coalfields and Victoria Falls) to agree to extend their line via Kansanshi to a point on the Katanga frontier within three years. They would have to agree to carry all supplies and copper for no more than £3 3*s*. 0*d*. per ton between the frontier and any Portuguese ports, and persuade the Wankie Colliery Company to supply up to 500 tons of coke per month for the mines. If Chartered paid Tanks the £250,000 expenses incurred in connection with the Benguela concession and agreed to guarantee the subscription of £3,000,000 of the £4,000,000 debentures the Central African Corporation was to issue, Rhodesia Railways would be able to connect to the Benguela Railway which, Williams claimed, would make Lobito Bay the natural port for all passenger and light goods traffic for the Rand and elsewhere in southern Africa. Rhodesia Railways could also form a company to develop the mineral grants in the concession. These proposals were put forward simply as a basis for discussion,

[11] Minutes of Meeting, 6 May 1903, TC 25.
[12] Gosselin to Lansdowne, No. 97, 17 May 1903, FO 2/769.
[13] Gosselin to Lansdowne, No. 103 Africa, 27 May 1903, FO 179/385.
[14] Machado to BRC London Directors, 1 June 1903, and BRC Board Minutes, 27 May 1903, TC 25.

being subject to the approval of Tanks's directors and shareholders, the CSK, the Benguela Railway Company, and the Portuguese government.[15]

Chartered completely reversed what had appeared to be a more co-operative attitude, apparently under pressure from Beit, and now said that the only basis for discussion was the conditions imposed in 1901. Chartered would agree to bring their railway to the Katanga border within five years of its reaching the Zambezi River in exchange for a half-interest in Katanga minerals.[16] Leyland was afraid that Chartered was being pushed by 'the Powers' to secure a north–south connection with the Benguela Railway, but that they were not at all concerned about the terms on which this might be accomplished. In addition, Lady Warwick's 'draining the life blood out of [their] capital' made it impossible for them to play a 'waiting game'.[17] Lady Warwick at this time found herself increasingly in need of money. The £150,000 shares Beit had not taken up in 1901 were still on her hands. Leyland took an option on them but was not happy about trying to find large amounts of cash at this particular time. In August 1902 he had taken up £5,000 which enabled the Countess to repay part of the £16,000 she owed her brother-in-law, the Duke of Sutherland.[18] By the end of the following May he had provided a total of £33,417.[19] The Countess might very well have been able to dispose of her shares on the open market, but this was certainly something Leyland and Williams would have wanted to prevent. In selling to Leyland under option, she also had the advantage of a fixed price.

By the end of June Leyland completely lost patience and drew up a biting catalogue of Williams's 'sins', the most serious of which was that Leyland had been led to believe that the CSK would put up 60 per cent of the money for the Benguela Railway but did not want their name to appear publicly in connection with it. Now it seemed that they would not participate at all. More specifically Leyland complained that the £100,000 he had advanced for the deposit with the Portuguese government and which he wanted to use to take up

[15] Williams to J. F. Jones, by this time a director of RRL, 3 June 1903, copy in CO 417/379.

[16] Jones to Williams, private, 10 June 1903, copy in CO 417/379.

[17] Leyland to Williams, 14 June 1903, TC 25.

[18] Sutherland to Leyland, 11 Aug. 1902, TC 25.

[19] Scotland to Leyland, 25 Feb. 1903, and Leyland to Scotland, 1 June 1903, TC 25.

more of Lady Warwick's shares and debentures issued by Warwick Estates Ltd., had not been repaid within the stipulated three months. He felt that constantly postponed arrangements to build some 'castle in the air' were draining their resources and could only lead to a débâcle. Chartered had not modified their position in any way and would probably come in and take up the concession at 'wreckers' prices'. Leyland strongly urged Williams to take up a proposal which involved dropping the Benguela Railway altogether and refused to discuss the matter directly with Williams or any other Tanks director, referring anything related to the Central African Corporation to his solicitors.[20]

Rhodesia Railways now approached Williams with some new proposals which would have eliminated the need for Leyland to find the £2,500,000 he had underwritten.[21] They were prepared to extend their line to a point in Tanks's concession area in Northern Rhodesia within five years of reaching Victoria Falls in return for, *inter alia*, 50,000 Tanks shares.[22] This would have had some advantage for Kansanshi, and Williams told Droogmans that he might have to accept the offer, leaving the question of a link to Katanga to be settled later.[23] Despite the fact that the CSK continued to refuse permission for Williams to transfer any of his holdings the Tanks board now decided to go ahead with the printing of the draft prospectus of the Central African Corporation.[24]

Lady Warwick, hard pressed by creditors and finding Leyland unable to take up immediately the remainder of the shares on which he had an option, decided to take a more direct part in the affair. She told Leyland that she had discussed the situation with 'one of the best brains in the country' and could get an immediate interview with King Leopold. If Leyland would definitely agree to underwrite a portion of the debenture issue, she had 'reason to *know*' that Leopold would underwrite a large sum as well. She estimated that Leyland's shares of the guarantee would not have to exceed one million pounds, but that Leopold's participation would create such a market for the debentures in England that he would probably not have to put up any cash at all. She claimed to know how things had been 'muddled' in Brussels, and wanted to deal directly with Leopold, leaving

[20] Leyland to Williams, 23 June and 3 July 1903, TC 25.
[21] Williams to Leyland, 10 and 11 July 1903, TC 25.
[22] RRL to Williams, 14 July 1903, TC 25.
[23] Williams to Droogmans, 17 July 1903, TC 25.
[24] Extract from TCL Board Minutes of 21 July 1903, TC 25.

Williams out of the picture for the time being but armed with something definite from Leyland. 'Trust me,' she said, 'and we will win.'[25] The Countess arranged to see Leopold on 24 July[26] but their discussion, of which there is no record, was not fruitful.

What was behind Lady Warwick's certainty that she could arrange matters or with whom she had discussed them remain in the realm of pure speculation. On the same day that she wrote requesting an audience of Leopold, Droogmans once again wrote to Williams refusing authorization to connect the Katanga railway to the Rhodesian system.[27] As Leyland refused to provide any further finance unless the Chartered offer were accepted, Williams thought the Benguela Railway might have to be dropped entirely.[28] Many financial difficulties would have been avoided if it had been. Williams did not, however, immediately agree to the Chartered scheme. Leyland refused to sign the draft trust deed and prospectus of the Central African Corporation, as without a firm assurance from Leopold regarding the link to Katanga, security for debenture holders was inadequate.[29] Tanks decided to issue a circular to shareholders explaining the situation and to instruct their solicitors to institute appropriate legal action against Leyland for his refusal to honour his underwriting agreement.[30] By this time Leyland had put up a total of £125,000 in connection with the Benguela Railway while Zambesia Exploring owed him a further £5,000. He offered to accept Tanks debentures for the total amount and to take up £50,000 of the debentures himself, provided he was given sufficient time to find the money and that all claims against him were dropped.[31] Tanks did not think this adequate compensation and decided to proceed with the lawsuit and issuing the circular.[32] Leyland resigned his directorships.[33]

In January 1903 Williams had begun making arrangements through

[25] Lady Warwick to Leyland, 25 May 1903, and private, 21 July 1903, and Leyland to Lady Warwick, private, 25 May 1903, TC 25.

[26] Sir Hugh Gilzean Reid to E. Carton de Wiart, 22 July 1903, and Lady Warwick to [Leopold], 23 July 1903 and minutes thereon, Leopold II Papers III, BW 34, File of Correspondence—Lady Warwick.

[27] Droogmans to Williams, 23 July 1903, TC 25.

[28] Williams to Droogmans, telegram, 23 July 1903, TC 25.

[29] Rowcliffe, Rawle & Co. (Leyland's solicitors) to Ingle, Holmes, Sons & Pott (Tanks's solicitors), 27 July 1903, TC 25.

[30] Extract from TCL Board Minutes, 31 July 1903, TC 25.

[31] Rowcliffe, Rawle & Co. to Ingle, Holmes, Sons & Pott, 11 Aug. 1903, TC 25.

[32] Extract from TCL Board Minutes, 10 and 12 Aug. 1903, TC 25.

[33] Leyland to TCL and to ZEC, 13 Aug. 1903, TC 25. The case was taken to court, but was dropped when acceptable financial arrangements were made.

Sir Charles Metcalfe, who in conjunction with the engineering firm of Sir Douglas Fox and Partners was to serve as consulting engineer for the railway and George Pauling, the contractor who had built the Rhodesian railways, for preliminary surveys and preparatory construction work to begin.[34] By August it was obvious that this work had to stop as there was no money to pay for it.[35] In view of the fact that Leopold continued to oppose the link between the Katanga and Rhodesian rail systems the Tanks Board concluded that the Benguela concession was even more valuable than it would otherwise have been, apparently simply ignoring the fact that his equal lack of support for a link with that line made the concession almost completely worthless. Having resolved to do all they could to protect the concession they applied to the Portuguese government for permission to use half of the £100,000 deposit to pay for initial construction and surveys. The Lisbon Board of the railway company agreed to indemnify Tanks for any claims Leyland might make against them as that money still belonged to him.[36]

In putting their request the company said that the situation had arisen largely from the breakdown of arrangements that had already been made. Chartered, whose opposition to the Benguela line had been supported, they claimed, by the Rothschilds and other European and American financiers, was continuing to make demands for building the link to Katanga which were too exorbitant to be considered. (It would indeed have been rather strange if the Rothschilds had not supported Chartered in view of their financial backing for all of Rhodes's activities in southern Africa from the amalgamation of the diamond industry and the opening up of the Transvaal gold fields to the formation of Chartered itself.) When the time came for putting into effect arrangements Williams had previously concluded with Leopold for the linking of the Benguela and Katanga lines, the King did not directly disavow them, but severely criticized them and adjourned discussion of them. At this point Williams advanced the view that Leopold had been forced to do this in order to secure German support at a second Berlin Conference which some people were demanding as a result of the growing anti-Congo agitation in England and elsewhere. In view of the German desire for a railway

[34] Copies of relevant correspondence between January and June 1903 are in TC 25. Freeman, Fox and Partners, successors to Sir Douglas Fox and Partners, still serve the Benguela Railway as consulting engineers.

[35] TCL to Learmonth, telegram, 12 Aug. 1903, TC 25.

[36] Extract from TCL Board Minutes, 25 Sept. 1903, TC 25.

connecting the Transvaal with South West Africa, Williams thought checkmating the Benguela scheme might have been the price the King had had to pay. While there may conceivably have been some element of truth in this view, it is unlikely at best; the settlement of the 'Nile question' was clearly of greatest importance.[37]

The recently changed Portuguese government was willing to make things slightly easier for the railway company, but was afraid that Williams might turn out to be nothing more than a speculator, whatever his reputation as a mining engineer. They did agree to extend the time limits for the construction of the different sections of the line, but would not allow any of the deposit to be used as they were afraid this might lead to difficulties with Leyland. This did little to help the railway's finances, and Pinto Basto threatened to resign unless some funds were forthcoming. He and Machado both decided, however, to remain in their posts, which Gosselin thought they would not have done unless advised to do so by the government. Gosselin was now in a much better position to understand the various aspects of the situation. The Foreign Minister, Wenceslau de Lima, discussed it with him, continuing to think Williams had had Gosselin's support originally despite reminders that the British government had had nothing whatever to do with introducing Williams to the Portuguese. With the reverses suffered in the course of events, Williams began actively seeking the official support he had previously so assiduously avoided. In November 1903 he and Lane made the first of many visits to the Legation to discuss the railway company's position.[38]

Despite the apparently increased confidence indicated by Pinto Basto's decision not to resign, by March 1904 nothing concrete had been accomplished in the way of finance. Williams strongly denied that there was any truth in a Lisbon newspaper report that he had severed his connection with the railway company. Lane pointed out that the material for a bridge across the Catumbella River had been ordered and that work would be going ahead on it shortly. Negotiations with Pauling were continuing, he claimed, and a contract was about to be signed.[39] This last was bravado; talks with Pauling were indeed going on, but only on a very tentative basis. The basic problem

[37] Gosselin to Lansdowne, No. 158 Africa, confidential, 19 Oct. 1903, FO 179/385.

[38] Ibid., No. 160 Africa, confidential, 24 Oct. 1903; No. 166 Africa, 1 Nov. 1903 and No. 190, very confidential, 27 Nov. 1903, FO 179/385.

[39] Gosselin to Lansdowne, No. 69 Africa, 31 Mar. 1904, FO 179/397.

remained: Leyland refused to finance the line unless Leopold changed his attitude. Leopold was adamant. If anything were to be accomplished at all, an entirely new source of finance had to be found.

It is only possible to piece together a superficial sketch of how this new source of funds was found. A stockbroker, Frank Hilder, was impressed by the fact that throughout the railway concession's difficulties, Williams had retained his faith in Tanks's future and had not disposed of any of his shares in it. Hilder introduced Williams to Athol Thorne, a financier, and Cecil Budd (later Sir Cecil) who was connected with the metal brokers, Vivian Younger and Bond, and was to become an important figure in determining British policy on non-ferrous metals during and after the First World War, setting up, at government request, the British Metals Corporation. With Lord Howard de Walden whom Williams had known in Africa and who was a friend of Lady Warwick, these men raised the money needed to start construction and later participated in development of the Katanga mines. They arranged for Mr. (later Sir) John Norton Griffiths, an associate of Thorne, rather than Pauling to begin the work, despite the fact that Griffiths had never done this kind of work before.[40] By the end of October 1904 arrangements were sufficiently advanced for Williams to be able to tell Sir Francis Bertie, Assistant Under-Secretary at the Foreign Office, that there was no longer any financial problem.[41]

LABOUR PROBLEMS

Construction of the first section of the line had to be carried out over very difficult terrain. From Benguela at 7 metres above sea level to Portela, a distance of only 60 kilometres, the land rises steeply to 902 metres, dropping again to 540 metres over the next 26 kilometres, rising again to 910 at kilometre 197 where Griffiths stopped work in 1908. From that point the geographical obstacles were slight; the terrain continues to rise, but much more gradually, reaching its highest point of 1,851 metres at kilometre 477. From there it slopes gradually downwards to 991 metres above sea level at Dilolo, just over the Congo border, some 1,346 kilometres from Lobito. Parts of the first sections were so steep that a rack and pinion section of

[40] Various memoranda in the Williams Papers, Hutchinson and Martelli, pp. 152–3, and information from Gwladys, Lady Norton Griffiths.

[41] Williams to Machado, private, 25 Oct. 1904. Williams Papers.

track had to be installed over about 8 kilometres, which was only converted to normal track by extensive realignment in 1948.[42]

In addition to the financial and geographical obstacles to construction of the line, there was also a serious shortage of labour. This might seem surprising in view of the functioning of the *serviçâo* system in Angola which was notorious for providing labour for the cocoa plantation on São Thomé and Principe, under recruiting and working conditions that aroused the anger of many people, particularly in Britain. The Benguela Railway Company was not supplied with any labour from any source by the Angola administration at this time. The company and Griffiths were forced to do their own recruiting and had to look fairly far afield. In 1905 agents in the Gold Coast, for example, sought permission to take 120 men to work on the line. There was some feeling in the British Colonial Office that this should not be allowed on account of the adverse reports that had been received as to the treatment of Africans in Angola and because of the restrictions on the removal of labour imposed in 1898 by the Governor of Lagos, Sir William MacGregor. It was pointed out, however, that MacGregor's action had not been based on any allegations of ill treatment, but on his convictions that these men could not be spared from Lagos. In any case, the railway company itself had not been attacked for its treatment of workers, and allowing the recruits to leave would in no way affect the question of recruiting for South Africa, the point with which MacGregor had been specifically concerned. The presence in Angola of Michael Holland, who was then acting as Williams's representative on the construction works, and who was to supervise the labourers' conditions, was a strong point in favour of allowing the men to go. Permission was granted and further recruiting allowed in 1907.[43] Some men from Sierra Leone and the Gold Coast also reached the works by way of the Congo where they had originally gone to work on the railway from Matadi to Leopoldville. They had subsequently become distressed British subjects and chose to accept employment in Angola instead of

[42] BR 39 and 40, *passim*; W. A. Hance and I. S. van Dongen, 'The Port of Lobito and the Benguela Railway', *Geographical Review*, xlvi, No. 4 (1956), 469, and H. F. Varian, 'Geography of the Benguela Railway', *Geographical Journal*, lxxviii (1931), 497.

[43] H. Bryan, Colonial Secretary and Acting Governor of the Gold Coast, to CO, telegram, 2 Oct. 1905 and minutes thereon, CO 96/432 and A. W. Ricketts, Griffiths's recruiting agent, to CO, 2 Mar. 1907, CO 96/465.

immediate repatriation.[44] It remained necessary, however, to find a single large body of recruits to keep work going according to schedule.

One idea was to recruit in Moçambique, but the Portuguese government was unable to provide any practical assistance because of the existing arrangements for recruiting Moçambique workers for the Transvaal mines. Additional recruiting for the railway would, it was felt, have serious repercussions for commercial and agricultural interests in Moçambique itself. The Natal government allowed Griffiths to recruit up to 2,000 Indians from among those who had originally gone there to work on railway construction. By mid-February when 1,000 recruits were ready to embark the Natal authorities refused to permit their departure without specific authorization from the government of India.[45] This the government of India refused, as existing legislation only gave control over free Indians embarking from Indian ports. Unless the recruits were indentured, which the India Office assumed they were not, they could not intervene.[46] Why this assumption was made is incomprehensible; if the men were not indentured, there was no reason for the Natal government to have raised the question of authorization at all. Someone in the India Office should have thought at least to raise the question of the recruits' status; no one did. Natal pressed the government of India to reconsider, stressing that Griffiths's agents had a good reputation and gave what the Indian Immigration Board in Natal regarded as sufficient guarantees as to working conditions and provisions for repatriation. The Natal Protector of Indian Immigrants agreed to look after the men's interest in Angola, and allowing them to go would be to their own advantage as they would otherwise be unemployed.[47] The Viceroy acquiesced.[48]

With Soveral's support, the company also asked to be allowed to recruit directly in India.[49] The British nature of the undertaking gained the request some Foreign Office support, but there was con-

[44] Arthur Nightingale, British Consul in Loanda, to FO, No. 21 Africa, copy in FO to CO No. 13921, 6 May 1907, CO 96/462.

[45] Griffiths & Co. to FO, most urgent, 21 Feb. 1907, IO File. All IO references are to J and P File No. 779.

[46] Telegram from Viceroy, 24 Feb. 1907 and minutes thereon, IO File.

[47] Sir Henry McCallum, Governor of Natal, to CO, telegram, 7 Mar. 1907. CO 179/240.

[48] Telegram to Viceroy, 18 Mar. 1907, and telegram from Viceroy, 23 Mar. 1907, IO File.

[49] IO Minutes, 15 Mar. 1907, IO File.

siderable discussion of the way in which the Portuguese were reputed to treat workers. At the same time, there was thought to be still some force in the old argument that the sooner the interior was opened to 'civilization' the sooner the slave-trade could be effectively controlled. Unofficially the India Office was quite disturbed by the idea, particularly by the revelations made by H. W. Nevinson in his book, *A Modern Slavery*, first published in 1906.[50] Officially the India Office merely said that direct recruitment in India could only be permitted on the basis of a convention reached between Portugal and the government of India. If the Portuguese government wished to reach such an agreement, they would have to go through the ordinary diplomatic channels.[51] The matter was dropped.

H. G. A. Mackie, British Consul at Loanda, reported on his general observations of the labour situation on the railway works and on his specific investigations of the treatment the Indians received. He felt initially that if careful records were kept there was no reason for apprehension. Climatic conditions varied from one area to another. In the generally unhealthy Benguela District, high mortality could be expected, but the line had already been completed through the driest territory, the so-called 'Hungry Country'. The actual treatment the Indians received depended not on the administration but on the contractors in whom Mackie had more confidence than he had in Portuguese officials. Initial difficulties were caused by the fact that the contractors were not given sufficient warning of the arrival of 2,000 Indians and had not been able to prepare adequate accommodation for them. The Indians struck but were brought back to work with the help of Portuguese authorities. Mackie provided no details of the nature of this help, but thought the Portuguese would be prepared to give assurances that British subjects would not be subject to Angola labour law.[52] In the event, none of the anticipated problems regarding treatment of the Indians arose. Health conditions were a different matter, with outbreaks of beriberi and sleeping sickness taking a heavy toll.[53]

Claims from Natal that bad treatment had been responsible for many deaths and desertions were not substantiated by Mackie's

[50] Soveral to Grey, 25 Mar. 1907 and minutes thereon; Seton (IO) to Barrington (FO), 18 Apr. 1907, FO 367/48/8896; Seton to Sir Charles Lyall, Secretary, IO Judicial & Public Department, 11 Apr. 1907, IO File.

[51] IO to de Leao, Benguela Railway Co. representative, 2 May 1907, IO File.

[52] Mackie to Grey, No. 6 Africa, 28 June 1907, FO 367/48/8896.

[53] Mackie to Grey, No. 6 Africa, 26 July 1907, FO 367/48/8896.

investigations. He found hospital conditions much improved over what they had been earlier, with good, plentiful rations provided. The deaths specifically complained about had occurred not because of bad conditions but because of disease and attempts by deserters to cross the 'Hungry Country'.[54] Deserters were thought to be prompted not so much by the treatment they had received or by health problems but rather by their desire to escape debts owed in Natal. Mackie suggested that it would help to get some deserters back to work if Natal would waive arrears of taxes over £5.[55] The Natal government agreed to this[56] but it is most unlikely that news of this reached Angola in time to have very much affect on the labour situation.

By June 1908 Griffiths reported that the last of the recruits had been repatriated either to Natal or to India.[57] This had become a source of considerable argument. It had been found that some of the men had in fact gone to India under indentures. Any who did not hold certificates of domicile in Natal were sent directly to India, whatever their own wishes in the matter.[58] According to the indentured men's contracts, Griffiths was required to repatriate them only if they gave eight months' notice of their intention to return to India and contributed £6 towards the cost of the journey. The India Office considered this entirely unsatisfactory and urged that unless these conditions were changed, Natal should be required to take the men back. It was rightly assumed that Natal would not want to do this, and the question was amicably settled, but for some months deserters continued to show up in Benguela seeking repatriation as distressed British subjects.[59] By this time, however, the railway company was no longer directly involved in the matter.

The first section of the line to Monte Sahoa, 17 kilometres from Lobito, was completed in August 1906.[60] Twelve months later it appeared to at least one disinterested observer that the line would be

[54] Mackie to Grey, No. 14 Africa, 31 Dec. 1907, FO 367/48/8896.

[55] Ibid., No. 9 Africa, 6 Sept. 1907, FO 367/48/8896.

[56] Governor Nathan to Lord Elgin, No. 27, 4 Mar. 1908, CO 179/244.

[57] Griffiths & Co. to Lord Crewe, 19 May 1908, CO 179/231.

[58] CO to IO, No. 13049, 14 May 1907, enclosing copies of relevant contracts, IO File.

[59] Minutes on FO to IO, No. 33466, 14 Oct. 1907, CO to IO, No. 43369, 5 Dec. 1908, IO File; Nathan to Elgin, No. 27, 4 Mar. 1908, CO 179/244; IO to FO, 26 Nov. 1908, FO 367/88/4678.

[60] Mackie to Grey, No. 2 Africa, 4 Mar. 1907, FO 367/48/1192.

finished by 1912.[61] Technically this would have been possible, but after Griffiths reached kilometre 197 at Cubal in 1908, financial and political obstacles arose to delay completion. In the meantime, however, new important developments had been coming to fruition in the long struggle to arrange a link between Katanga and Northern Rhodesia.

[61] Lt. C. K. MacLean, Commander HMS *Dwarf*, to Admiralty, No. 16, 20 Aug. 1907, copy in Admiralty to FO No. MO1278, confidential, 2 Oct . 1907, FO 367/48/11692.

CHAPTER IV

The Link to Beira

LEOPOLD CONSOLIDATES

Leopold was not in fact opposed to the Benguela Railway itself. He agreed that it would provide the most direct economical transport route for Katanga traffic, but was not, however, motivated solely by economic considerations. In addition to wanting to use Williams in his attempts to secure an outlet to the Nile, he also wanted to ensure that Katanga would become an integral part of the Congo State rather than an annexe, political or economic, of neighbouring British territory. Fears on this score were far from groundless. Apart from the geographical and cultural ties that linked Katanga more closely to Northern Rhodesia than to the rest of the Congo, the involvement of British finance and the fact that many of the people who first went to Katanga were from South Africa and the Rhodesias, meant that Katanga could easily have been drawn into the British orbit. Once the last doubts about the value of the mineral deposits had been dispelled, Leopold set about strengthening his position.

The first step was taken by van Eetvelde who proposed the building of a railway from Katanga to Matadi entirely through Congolese territory. The fact that this would be less economical than the Benguela line was well compensated for by the political advantages to be gained and by the opening up of the Kasai region. Van Eetvelde also suggested that Williams's Angola interests and some of his Katanga holdings be amalgamated with the CFK, a plan very similar to Williams's ideas for the Central African Corporation. This was despite some doubts van Eetvelde had about Williams's ability to participate in any new schemes or indeed to meet his existing financial commitments—doubts which Chamberlain had encouraged early in 1903.[1]

In April 1904 Leopold approved these plans and asked Williams to take responsibility for prospecting in Kasai and making a survey

[1] Van Eetvelde to Leopold, 16 Feb. and 13 Apr. 1904, VEP 35; 'Note pour le Roi—Chemin de fer au Katanga', 17 Dec. 1900, by van Eetvelde, EIC, IRCB 506.

of the proposed railway. The Congo State and Williams would share the costs of preparatory work equally, and a new company formed to subscribe or find subscribers for up to £6,000,000 to build the railway and exploit any valuable minerals that might be found. Williams was attracted by the idea, but refused to commit himself to such a large expenditure until he knew the value of any mines in the region.[2] He delayed a definite decision, most probably, as the Belgians believed, to see what would happen regarding Benguela Railway finance.[3] It was only in June that he agreed to undertake the prospecting, but did not think there was enough time to carry out the railway survey properly.[4] After reaching basic agreement with Leopold[5] he cabled George Grey to return home on leave via Boma, studying the Kasai region on the way. By the time the cable arrived in Katanga, however, Grey had already left via Bulawayo. Williams withdrew from the arrangements as there was no one else he trusted enough to send in Grey's place.[6]

The building of the railway line was later undertaken by the Compagnie du Chemin de fer du Bas-Congo au Katanga (BCK) which was one of the three large companies formed in 1906 to develop the resources of the southern Congo. Williams did not participate in this company or in the Société Internationale Forestière et Minière (Forminière) also founded in 1906 to exploit resources of Kasai with the assistance of American and British capital. Williams's non-participation in these two companies cost him not only a share of the profits ultimately gained from the development of Kasai but also some of the goodwill of his Belgian colleagues. Forminière became closely associated with the Companhia de Diamantes de Angola (Diamang) which had interests somewhat opposed to those of the Benguela Railway.

In May 1905 Leopold, who was taking the waters at Ems, Germany, summoned Williams for a discussion of their mutual interests. Williams provided a 'High Authority' explanation for this summons, claiming that on Christmas Day 1904, in conversation with Gosselin in Lisbon, he had suggested that matters might be resolved if Kings

[2] Droogmans to Leopold, 5 and 7 Mar. 1904, EIC, RF 1713.

[3] N. Arnold, EIC Director of Finance, to Leopold, 7 July 1904, enclosing a copy of Empis to van Eetvelde, 2 July 1904, EIC, RF 1713.

[4] Williams to Arnold, 16 June 1904, EIC, RF 1713.

[5] Minute by Leopold, 10 Sept. 1904, on Arnold to Leopold, 18 Aug. 1904, EIC, RF 1713.

[6] Williams to Arnold, 19 Sept. and 19 Dec. 1904, EIC, RF 1713.

Edward and Leopold met to discuss them. Williams thought that it was as a result of this suggestion that the two monarchs met at the theatre in Paris a short time later and worked out his railway problems at the dinner they were supposed to have had together after the performance. King Edward may indeed have tried to help Williams at this time, but there is no evidence that Gosselin ever acted on his suggestion. The royal meeting and dinner were widely reported in the Paris press at the time, but according to one of Edward's private secretaries, Sir Frederick Ponsonby, the encounter at the theatre was only a matter of bad luck, and the dinner never took place. Ponsonby maintained that Leopold was well aware of his English cousin's dislike for him, and had kept his distance while they were both in Paris.[7] In any case, the immediate outcome of the talks at Ems hardly showed that Williams's problems had been settled.

Leopold continued not to agree to the connection between the Benguela Railway and the mines, but did put forward a variant of van Eetvelde's amalgamation scheme. The discussion was primarily concerned with the fundamental problem of bringing the mines to production. George Grey had been supervising the further prospecting of the concession area, but nothing concrete had been done about forming the exploiting companies mentioned in the 1900 Convention. Leopold now proposed the formation of a Belgian company with a capital of £100,000 to take over both the CSK's and Tanks's mineral holdings in Katanga and work the mines.[8] When this was later discussed by the CSK, Grey, who could hardly have been speaking without Williams's approval, supported the idea of a single company rather than the several originally envisaged.[9] There were certainly sound economic justifications for this, but the fact that no companies to exploit Katanga's wealth were registered in London helped to weaken Williams's position in Katanga. It is highly possible that agreeing to this was the price he had to pay to secure any agreement at all. On 28 October 1906 the third of the large companies formed in that year, Union Minière du Haut-Katanga was established by royal decree. In addition to the right to exploit minerals, this new company could establish any commercial, agricultural, or railway enterprises which would contribute to Katanga's economic development. Baron

[7] Undated memorandum by Williams, 'Negotiations in connection with Pooling and other railway agreements', Williams Papers; F. Ponsonby, *Recollections of Three Reigns*, 2 vols. (Eyre & Spottiswoode, London, 1951), ii. 222.

[8] Leopold to van Eetvelde, 2 June 1905, VEP 19.

[9] Private information.

Ferdinand Baeyens, Governor of the Société Générale de Belgique became Union Minière's first president, Williams its first vice-president.[10]

The following year saw more concentrated efforts being made to develop the copper deposits, the gold mine at Ruwe, and possible power sources. All technical work was under the control and direction of Robert Williams & Co. With the prospect of copper being available shortly, the need for a railway link to the coast became even more urgent. As neither the all-Congo line nor the Benguela Railway could be completed very rapidly, the Rhodesian rail system, which had reached Kalomo just beyond the Victoria Falls in 1904, provided the best potential solution to the immediate problem. Chartered's demands were no longer as extravagant as they had been, but before agreeing to extend their line to the frontier they wanted Union Minière to agree to pay ¾*d.* per ton/mile from Beira to Bulawayo, dropping to ½*d.* per ton/mile only when traffic surpassed 1,000,000 tons annually. Construction material was to be charged at the rate of 1*d.* per ton/mile in full train loads. North of Bulawayo Chartered wanted to charge actual cost, including the cost of servicing debentures secured on that portion of the line as well as operation expenses, subject to a maximum of 1½*d.* per ton/mile.[11] These terms were entirely unacceptable to Union Minière who did not, however, want to put too much pressure on Chartered for fear of bringing the negotiations to an end. They authorized Williams to accept any terms he saw fit provided that two conditions were met: Union Minière was to have the right to deal subsequently with any other railway that might become available to them, and Chartered had to agree to offer them any lower rates they might later offer anyone else.[12] On 20 December 1906 Williams told Jameson, with whom he had been negotiating informally, that he accepted the rates they had agreed.[13] Rhodesia Railways officially disclaimed any knowledge of talks between the two men but at the same time said that the rates proposed —as a result of those talks—had not been intended to apply to copper and tin.[14]

Williams was convinced that Chartered were being difficult because of the presence on their Board of some people inimical to the Katanga

[10] *Union Minière du Haut-Katanga . . .*, pp. 67–73.
[11] RRL to Williams, 18 Oct. 1906 and 28 Jan. 1907, Williams Papers.
[12] UMHK to Williams, 13 and 18 Dec. 1906, UM 76/345.
[13] Williams to Jameson, telegram, 20 Dec. 1906, Williams Papers.
[14] RRL to Williams, 28 Jan. 1907, Williams Papers.

interests.[15] One of the people he clearly had in mind was his Angola adversary, Edmund Davis, who was at the centre of some of the serious financial difficulties Chartered was facing at the time. As the leading member of the syndicate trying to exploit the zinc and lead deposits at Broken Hill, Davis had convinced the Mashonaland Railway Co. to continue the line from Kalomo. On the basis of Davis's assurances that by the time the line reached Broken Hill between 12,000 and 16,000 tons of ore would be ready for shipment to the coast, and that a steady flow of traffic, sufficient to cover costs, would be available, Chartered had agreed to guarantee the debentures issued to finance the extension. Davis's syndicate, which was later reorganized into the Broken Hill Development Corporation, had no liability whatever for the railway. It had not been required to give a prior guarantee of debenture interest, while neither the railway company nor Chartered had any direct control over its activities. In the event, Davis's assurances proved worthless. Technical difficulties encountered in trying to separate zinc and lead, which were found together over a large portion of the field, meant that traffic was far below expectation, producing grossly inadequate working revenue for the railway, and forcing Chartered to meet debenture servicing costs at the same time as they were seriously burdened by their guarantee of the debentures secured on the Beira lines.[16]

Earl Grey, who continued to take an interest in the company's affairs while Governor-General of Canada, urged Chartered to push the line on to Katanga in any case. This could be done, he pointed out, in a matter of months at a cost which was a mere 'bagatelle' compared with that of the Benguela line, the money for which he thought should be saved.[17] Grey did not explain why he thought Williams would be willing to abandon the Benguela Railway as he had finally got construction started, or how Chartered was to find even the small amount of money needed to put his idea into effect.

It was obvious that at this stage Union Minière and Chartered needed one another. On the one hand, Union Minière needed rail transport to develop its assets, and the link from the south could be completed more quickly than any other. Also, no deposits of coal

[15] Williams to Heyvaert, 8 Dec. 1906, Williams Papers.

[16] Henry Birchenough to P. Lyttelton Gell (both Chartered directors), undated, and Gell to D. E. Brodie, BSAC Assistant Secretary, confidential, 31 Aug. 1907, Gell Papers.

[17] Grey to Gell, 31 May and 25 Sept. 1907, Gell Papers.

suitable for making metallurgical coke had been found in Katanga.[18] This meant that coke would have to be imported from Europe or from Wankie, the latter being a much less expensive source of supply. At the same time, Union Minière had something of an advantage as Chartered's straitened finances meant that the link to Katanga was extremely urgent for them since there was no other source of substantial traffic available to them. Williams was impatient, but Leopold and his Belgian backers were in a better position to play their waiting game until they could secure advantageous terms.

While Chartered was trying to decide what to do, Williams and the Belgians were resolving some of their conflicting railway interests. The agreement reached was embodied in a 'Pooling Agreement' between the Benguela Railway Company, the BCK, and the CFK. The Benguela Railway Company agreed to bring its line to the Congo frontier at a point between 11° and 12° south latitude, thus ensuring that a considerable portion of the connecting line to the copper mines would pass through Congolese territory and bring some advantages to the western Katanga mining area. The BCK undertook to connect the line at the frontier to Ruwe where it was expected to meet the CFK coming north. Once this link was completed, the three companies were to pool their gross working receipts and divide them among themselves in proportion to the length of line each operated. Because of the high construction costs involved in the first section of the Benguela Railway, the length of its line was to be increased by 200 kilometres for purposes of calculating its share of receipts. The agreement was subject to revision in 1920 and every ten years thereafter, with six months notice required of any intention to modify or cancel it. The CFK was to build a line from the Rhodesia–Katanga frontier to the Étoile du Congo mine, the centre of initial operations, thence to Kambove and Ruwe.

Under the terms of this agreement if the BCK had ever been completed to Matadi the company would have shown greater profits if all copper were carried by the Benguela Railway than if the BCK had carried some of it themselves. This may or may not have been intentional, but as the Pooling Agreement never in fact went into effect it was of no practical importance. The signing of the agreement was the high point of co-operation between Williams and his Belgian

[18] Memorandum by H. Wilson Fox, Chartered's London Manager, 'Possibilities of Traffic from Tanganyika Concessions', 23 Oct. 1907 in: 'Letters and memoranda . . . in regard to . . . Rhodesian Railway problems', copy in Gell Papers.

colleagues. The financial and political interests of all concerned seemed assured. Dark clouds did not appear on the horizon until the eve of the 1914–18 War.

CHARTERED CAPITULATES

Chartered began the last round of negotiations for the extension of their line under the erroneous impression that the Belgians would be prepared to help them finance it. Philip Lyttelton Gell, a close friend of Lord Milner and Secretary to the Delegates of the Oxford University Press who had become a Chartered Director in 1899, argued that the completion of the link to the south had become a matter of Congo government policy; unless the Congo immediately had an extensive mining industry, they would be 'saddled' with a serious annual deficit. He also stressed Union Minière's need for coke from Wankie, and wanted Chartered and Williams to form a new company which would acquire the Kalomo–Broken Hill section of the Rhodesian rail system and continue it to Kansanshi or some other point to be agreed with the Belgians. Material for the extension would be ordered in Belgium if Belgian banks underwrote an issue of £2,500,000 3½ per cent bonds for the acquisition of the Kalomo–Broken Hill line.[19] This proposal would have relieved Chartered of a substantial part of her financial burden, but the Belgians had no reason to help them do so. Chartered finally agreed to build the line on terms very favourable to the Katanga interests.

On 11 July 1908 in Brussels—three days later in London—the BCK, CFK, Chartered, and Union Minière signed an agreement whereby Chartered was to form, by the end of January 1909, a company to build a railway from Broken Hill to a point on the Katanga border. From that point the BCK, on behalf of the CFK, would construct the line to the Étoile du Congo mine. The BCK would not have to begin construction until the Rhodesia line was completed. Neither Union Minière nor the BCK had any financial liability for construction in Rhodesia for which Chartered apparently expected the Alfred Beit Railway Trust to provide the finance. This trust had

[19] Confidential Memorandum by Gell, 22 Feb. 1908, in BSAC Commercial Minutes, 27 Aug. 1908, Gell Papers. In the aftermath of the Jameson Raid, Chartered was required to keep the CO informed of its administrative activities. Chartered then divided the agenda and minutes of its Board meetings into administrative and commercial sections, forwarding copies only of the former to the CO. The Chartered minutes in the CO files therefore give a picture of only a part—and in general the least interesting part—of the company's activities.

been established and endowed under the terms of Beit's will, and over the years played an important role in Rhodesia's economic development. In this case, however, it did not live up to expectations. Jameson urged Williams to participate fully in the negotiations with the Trustees, concentrating his efforts particularly on Bourchier Hawksley, Chartered's solicitor and one of the least co-operative and most influential of them.[20] Williams preferred to leave any role he might have in these talks, which he maintained did not concern him directly, to his solicitor. He did not want to take a sufficiently active part in them to make it possible for Tanks's shareholders to attach any blame to him in the event of their failure.[21]

Hawksley was disturbed by the fact that Chartered would have to spend some £750,000 before the BCK would have to spend anything. He tried to arrange for permission to borrow the £300,000 earmarked by the Belgians for the Katanga section to begin work in Rhodesia.[22] Jadot and the BCK refused to consider the idea; as the money had been secured as part of a large state loan specifically intended for developing means of communication in the Congo, it could not be used in any other way. Assuming that Beit would have wanted the extension completed had he still been alive, Jadot felt that it was up to Chartered to find their own money.[23] Williams, continuing to feign indifference, said he could not assist in any way as he still regarded the Benguela line as the best one for Katanga and advised Jameson to try to get support from the British government. Through his connections with the Grey family, Williams made his own soundings on this point. Sir Edward did not think there was much hope of getting any financial help from the government for the project[24] but asked the Colonial Office to discuss the matter with Williams. Colonel Seely, Parliamentary Under-Secretary at the Colonial Office, confirmed Grey's view and emphasized that the link line when completed might very well prove entirely worthless because of the provisions of the British South Africa Company–Germany railway agreement of 1899.[25]

[20] Jameson to Williams, telegrams, 25 and 27 July 1908, Williams Papers.

[21] S. Pott (of Ingle, Holmes Sons & Pott) to Williams, telegram, 24 July 1908, Williams to Jameson, telegram, 24 July 1908, and Williams to Jadot, 27 July 1908, Williams Papers.

[22] Pott to Williams, 28 July 1908, Williams Papers.

[23] Jadot to Williams and BCK to Williams, 31 July 1908, Williams Papers.

[24] George Grey to Williams, 21 Sept. 1908, Williams Papers.

[25] Seely's minute on Williams to Seely, 4 July 1909, CO 417/480.

This agreement provided that Chartered's railway lines would not be connected to the west coast of Africa north of 14° south latitude unless a link had already been completed through South West Africa. Any westward extension south of 14° latitude had to pass through German territory. In March 1898 Chartered and the German government had reached agreement for the construction of the Cape-to-Cairo telegraph line through German East Africa. The Germans refused to ratify the agreement, however, unless the railway agreement was concluded at the same time.[26] For Chartered this was simply a means of securing the telegraph route and a possible step towards securing a route for the Cape-to-Cairo Railway. For the Germans the agreement was much more important as it represented a possible means of securing a rail connection between their territory and the Transvaal through the Otavi mine area. It was also another means of applying pressure on the Portuguese to force them to give way to German demands in southern Angola.[27]

The Colonial Office made quite a fuss about this point, expressing the fear that Germany would expect the British government who, although not a party to the agreement, had tacitly approved it, to force Chartered to honour its obligation to build a line to South West Africa. This would involve the government in the construction of a useless railway across Bechuanaland. Hawksley at first spoke rather off-handedly of removing a rail somewhere along the line to break the through connection, but under pressure officially informed Williams of the existence of the 1899 agreement and warned him not to try to raise money on the prospects of through traffic from Lobito to Beira. Williams, clearly relishing the position of strength he found himself in, replied that he wanted to be left alone to go about the business of finishing his own railway; Chartered's railway agreement with Germany had nothing to do with him. In any case he did not believe that Chartered intended by breaking the line to 'cut off its nose to spite its face'. The Colonial Office then officially reminded Chartered of their obligation under the Supplemental Charter of 1900 to keep the government informed of their administrative proceedings. It was now Chartered's turn to make protestations of innocence. Claiming to have had no knowledge of the 'Pooling Agreement' they maintained that a clause had been inserted in the

[26] Copies of relevant correspondence can be found in CO Confidential Print, No. 574 Africa under item No. 265.

[27] cf. Drechsler, *passim.*

July 1908 agreement to cover the possible eventuality of severing the link. The Belgians said the clause had been inserted at their request because they might want to break the connection in order to ensure full use of the BCK and Benguela Railways.[28]

The Colonial Office were convinced, quite reasonably, that neither the Benguela Railway Company and the BCK, nor Chartered would want to break any connections once the lines were completed. They were also afraid of becoming involved in complaints of fraud if Williams's shareholders, hoping to profit from through traffic, were not informed of the situation.[29] Lengthy discussions took place involving the Foreign Office, Chartered, and the Colonial Office. At this point Chartered maintained that the Benguela Railway would never be completed because of construction difficulties in the south-western Congo and the climate of the country through which the line would have to pass. If these obstacles were overcome, Chartered was sure the Benguela Railway would take all the Katanga traffic, making it possible for them to 'scrap their line' with perfect equanimity. The break would indeed be welcome as they would not want Rhodesian traffic to go via Lobito. Chartered could not have been arguing seriously; no mention of alternative sources of traffic for their railways or of the problem of coke for Katanga was made. The company very probably took the same view as the Foreign Office, which did not see eye to eye with the Colonial Office on many aspects of Anglo-German relations; the line should only be broken if and when the German government complained about it.[30] The Colonial Office finally settled the matter to its own satisfaction by asking Chartered for assurances that they would undertake to break the connection when asked to do so by the British government to prevent a breach of the 1899 agreement. Chartered readily agreed.[31]

The Beit Railway Trustees, perhaps more in keeping with what Beit would have wanted than Jadot had assumed, refused to finance the extension. Reflecting Jameson's hope that Williams would be able to do something to find the money 'later on',[32] Chartered

[28] CO Minutes, 11–20 Nov. 1908, attached to which are copies of relevant correspondence and agreements, CO 417/463; BSAC to CO, 21 May 1909, CO 417/476.

[29] Minutes on FO to CO, No. 25033, 2 July 1908, CO 417/473.

[30] Minutes on CO to FO, No. 17624, 2 July 1909, FO 367/130/6439.

[31] CO to BSAC, 24 Sept. 1909, CO 417/473, and BSAC to CO, 2 Nov. 1909, CO 417/476.

[32] Jameson to Williams, telegrams, 10 and 14 Sept. 1908, Williams Papers.

officially informed Williams that they could not find the funds to establish the company required under the July 1908 agreement. They did not think 'any useful purpose could be served', however, by allowing that agreement to lapse.[33] Williams's protestations of non-involvement in the matter had not prevented his having a complete proposal for fulfilling the agreement immediately to hand. He was prepared to form a company which would build the railway, as well as take over the working of Kansanshi and a further 10,000-square-mile land and mineral concession he wanted Chartered to give him. Chartered's interest in Kansanshi was to be reduced from 30 per cent to 10 per cent. As the idea still persisted that the main line of the Rhodesian rail system would go to Lake Tanganyika, Williams also insisted on being assured of every facility to connect the concession area to the main line. He hoped to be able to convince Pauling to agree to build the extension to Katanga for £750,000 and to take up £250,000 debentures himself.[34]

In September 1908 the Marquis of Winchester and Henry Birchenough, the Chartered directors Jameson considered most personally acceptable to Williams, were asked by the Chartered Board to carry on further negotiations. Without Hawksley's 'irritating presence' Jameson was sure some arrangement could be made.[35] After preliminary talks in England and in Belgium, the three men reached a tentative agreement that gave Williams virtually all he wanted. The talks in Brussels had confirmed Winchester's belief that the Belgians would not accept the terms originally put forward by the Board, but he felt the terms agreed would satisfy Chartered's obligations, and urged the Board to accept them.[36]

Gell was probably not alone in opposing acceptance, and appears to have been most reluctant to do anything that might be useful to Williams in any way. This attitude was, at least in part, a result of his belief that Williams was trying, or would try to 'squeeze' Rhodesia Railways.[37] This belief was undoubtedly well justified, but Williams was clearly in the stronger position and Chartered had little

[33] BSAC to Williams, 8 Sept. 1908, Williams Papers.

[34] Williams to Scotland, ? Sept. 1908, Williams Papers.

[35] Jameson to Williams, telegram, 17 Sept. 1908, Scotland to Williams, telegram, 17 Sept. 1908, Williams to Jadot, telegram, 18 Sept. 1908, and Williams to Scotland, 18 Sept. 1908, Williams Papers.

[36] Ingle, Holmes, Sons & Pott to Williams, 16 Oct. 1908, Williams Papers; Winchester to Gell and Fox to Gell, 3 Oct. 1908, Gell Papers.

[37] Winchester to Gell, 6 Oct. 1908, Gell Papers.

alternative but to co-operate with him. Gell's objections were overruled. Before any final arrangements could be made Williams had to secure finance. Pauling came up to expectations and Williams suggested that Union Minière acquire his 40 per cent share in the CFK, leaving him the option of repurchasing it in three or four years. This would free him for the time being of the responsibility for finding 40 per cent of the money to build the Katanga portion of the line and leave him unfettered in making the necessary arrangements for financing the Rhodesian section. Jadot greeted this suggestion with considerable enthusiasm.[38] Williams never regained his holding in the CFK but did remain its vice-president.

After some further discussion Chartered and Tanks signed an agreement whereby Tanks was to form and finance a company to satisfy the terms of the 1908 agreement. Chartered gave Tanks the right to construct and operate a railway from Broken Hill to a point on the Congo frontier, the right to connect Kansanshi to that line or to the CFK, and the right to build branch lines to any mineral areas discovered. Tanks was granted twenty 300-acre coal areas, mineral rights excluding precious stones in one or more 2,500 square mile areas, and the right to select, within two years, two 6,000-acre farms in north-western Rhodesia. Chartered retained a 15 per cent interest in the coal and mineral concessions, and after ten years, or any subsequent period of five years, could redeem the railway for the cost of construction plus 10 per cent. The new company was to have a maximum share capital of £1,600,000 and the power to issue up to £850,000 debentures. Four directors, including the Chairman, were to be named by Tanks, two by Chartered, and one by debenture holders. In return for Kansanshi and the railway concession the company would pay £400,000 in fully paid shares, £50,000 of which were to go to Chartered. Tanks guaranteed £750,000 fifty-year $5\frac{1}{2}$ per cent debentures, while Baron Émile d'Erlanger who had been Pauling's financial backer for many years agreed to complete the financing provided Williams could find £150,000 himself. This meant that d'Erlanger, in addition to whatever backing he may have been giving Pauling to take up his £250,000, was putting up £350,000 of his own. He imposed conditions which required some modification of the agreement, changes which were easily accepted. Slight alterations were made in the capital formation and the coal areas were

[38] Pauling to Williams, 5 Oct. 1908, Williams to Jadot, 26 Oct. 1908, and Jadot to Williams, 2 Nov. 1908, Williams Papers.

included in the assets to be turned over to the new company. D'Erlanger was named the company's banker, and had the right to appoint two directors.[39]

The Rhodesia–Katanga Junction Railway and Mineral Company (RHOKATS) was formed on 25 January 1909 and by agreement dated 13 May was formally recognized as the company envisaged in the July 1908 agreement.[40] Pauling signed the contract to build the railway in February and brought it to the frontier before the end of the year. Having previously arranged to participate in the Belgian firm constructing the CFK he continued on to the Étoile du Congo without interruption. The first locomotive from the south reached Elisabethville on 27 September 1910, only a few weeks after that town had become Katanga's administrative centre.[41]

Construction had begun in June 1909 of a smelting plant on the Lubumbashi River. This plant's water-jacket furnace was run off for the first time on 30 June 1911. A report called for by Union Minière in 1907, by a M. Frêcheville had predicted some production difficulties, but was ignored at the time because it did not agree with Williams's view of the situation. Frêcheville was apparently right, however, as technical problems made it impossible to begin full-scale operations for several months.[42] By this time considerable capital had been invested and Union Minière was no longer in a position to play a waiting game. A very serious financial situation was developing. Jadot and Dr. Sheffield Neave were sent at the head of an inquiry mission to investigate the problems and suggest solutions to them. Neave was a founding director of both Tanks and Union Minière and had been involved in some way, not entirely clear, in the abortive attempts made in 1895 and 1896 to float Lady Warwick's

[39] D'Erlanger & Co. to Pauling & Co., 13 Nov. 1908, copy in Pauling & Co. to Williams, 13 Nov. 1908, and copies of the relevant agreements, all found in the Williams Papers.

[40] BSAC to CO, 21 May 1909, CO 417/476.

[41] *Union Minière du Hart-Katanga*, pp. 94–5. In 1913 the Rhodesia–Katanga Railway was leased to the Mashonaland and Rhodesia Railway companies for an annual rent of £44,000 plus 5½ per cent of RhoKats's capital outlay and 50 per cent of the profits earned from working the line. ('Memorandum on the lease of the . . . Railway . . .' by J. B. G. Taylor, 23 July 1924, SA 241/4/1.) In 1928, Chartered, through the Mashonaland Railway Co., purchased the RhoKats line for a total of £975,000 (E. B. d'Erlanger, *The History of the Construction of the Rhodesian Transport System* (Burrup Mathisson, London, 1938), p. 29). RhoKats was reorganized as the Rhodesia–Katanga Co.

[42] A copy of Frécheville's report and much of the correspondence related to it are in the Williams Papers.

first concession. The investigation committee reported that the technical and administrative departments in Katanga, which had been kept entirely separate, had to be thoroughly reorganized to save the company from ruin. Neave was privately of the opinion that discord and friction between British and Belgian staff members was largely, though not entirely, responsible for the inability of the mining and smelting operations to produce copper efficiently. The most far-reaching aspect of the subsequent reorganization was the removal of the Technical Department from Williams's control. A Technical Committee, which included Williams among its members, was established in Brussels to exercise general control of policy. In Katanga the technical department was brought within Union Minière's own organizational framework. From the point of view of managerial efficiency this was unquestionably a good step; it also imposed further limits on Williams's scope for independent action. With the backing of the Société Générale, Union Minière was able to increase its capital by 2,500,000 francs and issue 10,000,000 francs debentures. This provided the funds to recover from the losses of the previous years and to expand production and the areas of exploitation. By 1913 the technical problems had been overcome and production was going smoothly. New mines had been opened up and reached by a branch railway line. That year 7,400 metric tons of copper were produced.[43]

One non-technical problem which added to the difficulties of production was the recruiting and maintenance of a sufficiently large unskilled labour force. In 1910 a non-profit-making limited liability company, the Bourse du Travail du Katanga, was established to supply labour for all Katanga employers. The Bourse never functioned very well and was unable to meet the heavy demands of Union Minière which, through Robert Williams & Co., had to recruit independently outside the Congo, primarily in Northern Rhodesia. This foreign recruiting continued until 1928 when Union Minière became the first large mining company in southern Africa to establish a stabilized labour force.[44]

[43] *Union Minière du Haut-Katanga*, pp. 96–109; Sheffield Neave Diary, 12 and 20 Dec. 1911. I have found no evidence to support the view privately expressed to me, originating from a source that cannot be considered unprejudiced, that Union Minière's financial difficulties in the early years were due to Buttgenbach's incompetence as the company's managing director.

[44] UM 75/346 and 347, *passim*; *Union Minière du Haut-Katanga*, pp. 163–5; Bruce S. Fetter, 'Elizabethville and Lubumbashi: the Segmentary Growth of a Colonial City, 1910–1945' (Univ. of Wisconsin Ph.D. Thesis, 1968), *passim*.

The smooth growth of production was brought to an abrupt halt when war was declared in 1914. All unnecessary development work was stopped, coal shipments from Wankie cancelled, labour recruiting halted, and materials already shipped but not immediately needed stored at Beira. Shipments of copper were halted and as many as possible of Union Minière's European employees transferred to the CFK. At Williams's suggestion, Union Minière moved its offices to London as it was certain that Britain would remain in contact with Africa, but might not be able to do so with Belgium. Williams and two other directors were authorized to maintain contact with the Belgian Colonial Minister and to take the decisions necessary to continue operations.[45] Production went on virtually without interruption during the war and new smelting furnaces were built. General development did not, however, keep pace with production, and this created serious technical problems that could only be solved after hostilities ceased.

[45] UMHK to Williams, Personal, 3 Sept. 1914, UMHK (Brussels) to UMHK (Elisabethville) telegrams, 3 and 19 Aug. 1914, and Williams to UMHK, 6 and 8 Aug. 1914, UM 64.

CHAPTER V

Construction Halted

GERMAN ATTEMPTS TO GAIN CONTROL

By the time Pauling stopped work on the Benguela Railway at Cubal in 1913, the security provided by the 'Pooling Agreement' was no longer sufficient to overcome the doubts about its future raised by its involvement in the complex Anglo–German relations that preceded the First World War. The aspect of these relations that most directly affected the railway was the negotiations for the revision of the 1898 'Secret' Treaty regarding the Portuguese colonies, which in turn was involved in the dream of a German *Mittelafrika*—German territory stretching across the continent from west to east. One of the foremost exponents of this dream was Wilhelm Solf who became German Colonial Minister in 1911, after having served as Governor of Samoa. Solf sought the help of German businessmen, but complained that their 'goodwill and patriotism . . . [were] immense . . . as long as there was no question of opening their wallets'.[1] There was some interest in the Congo among German financiers who wanted to see it developed as part of German East Africa's hinterland; by 1913, some 20,578,000 gold francs had been invested by Germans in Belgian companies operating in the Congo. Katanga was potentially a major source of traffic for the Central Railway from Dar-es-Salaam to Kigoma on Lake Tanganyika which was then nearing completion, but interest in Katanga had waned considerably when the initial technical difficulties revived fears that the area's mineral wealth could not be exploited profitably. In 1912, for example, the Deutsche Bank had only reluctantly taken up a concession of 1,235,000 acres in northern Katanga, and did nothing to develop it. Solf none the less considered Katanga as of major importance in his plans to secure control of the Benguela Railway for Germany.

The completed line, if controlled by Germany, would make available to them a transcontinental rail/water transport link. Belgian

[1] Solf to Rosen, 7 Nov. 1913, cited in Willequet, *Le Congo Belge et la Weltpolitik* (Université Libre de Bruxelles, Brussels, 1963), p. 391.

co-operation was essential for the completion and successful operation of the link; Williams's was not. Before taking any steps to acquire the Benguela Railway Solf wanted to secure the reversionary rights over the strip of central Angola assigned to Britain by the 1898 Treaty in order to make all the territory through which the railway was to run potentially German. Informal negotiations for the exchange of territory began in December 1911 between Solf and Lewis Harcourt, British Colonial Secretary, 1910–15. By the end of April 1912 tentative agreement was reached between the two men, Solf gaining the reversion he wanted. Neither Harcourt nor anyone else in the Colonial Office was aware of why Britain had been assigned the central strip in the first place, and when the Foreign Office, with traditionally more pro-French leanings than the Colonial Office, were informed of the terms agreed, they criticized them on more general grounds, not objecting to the exchange of territory itself. Grey, however, overruled all objections from his department, maintaining that informal negotiations, on matters that were 'really colonial', were desirable. Apart from the general wish—shared by Harcourt—to see an improvement in relations between Germany and Britain, which had been exacerbated by the Haldane mission, Grey was primarily concerned with getting the Germans to agree to the publication of the revised treaty, the 1898 treaty, and the 'Secret' Treaty of 1899 between Britain and Portugal (the so-called Windsor Treaty). He thought the territorial exchange was the price that had to be paid for publication, and was prepared to initial the revised treaty even before it was approved by the Cabinet. Cabinet approval came in June 1912 but delays on the German side postponed initialling until October 1913. Because of German reluctance to have the Anglo–Portuguese Treaty of 1899 made public, the revised Anglo–German Treaty had not been ratified by the time war broke out. Solf none the less operated on the assumption that the partition of the Portuguese colonies could be provoked and that the Benguela Railway would run through German territory.[2]

In May 1912 Williams was approached by a Continental group wanting to purchase all the Benguela Railway Company's debentures held by Tanks and Zambesia Exploring, and to find £2,000,000 to

[2] The preceding is based primarily on Willequet, pp. 379–99, and P. H. S. Hatton, 'Harcourt and Solf: the Search for an Anglo-German Understanding through Africa, 1912–1914', *European Studies Review*, i, No. 2 (1971), 123. Further references can be found in these works.

complete construction to the frontier. Williams claimed not to know exactly who was behind the offer, but the Foreign Office, whom he asked for advice, assumed that German interests were involved. At about the same time Baron Bruno Schröder and Frank Tiarks, who were to become a part of one of Williams's unsuccessful attempts to finance the railway after the war, were trying to establish a chartered company in Angola with German financial participation. The Foreign Office thought that, while the revised treaty was still under negotiation, it was best to tell Williams as little as possible about the extent of possible German interest in his railway.[3] Unofficially Grey told Williams the terms of the 1898 Treaty, which Williams already knew,[4] while officially the Foreign Office simply said that agreements with Germany made it impossible for His Majesty's Government to give political support to the Benguela Railway. Williams had to consider the approach from the German banking group solely on commercial grounds. If possible, they wanted him to postpone doing anything definite.[5] Grey subsequently went so far as to say that if it were absolutely necessary for purely commercial reasons to bring capital other than British or Portuguese into the railway, the Germans should be given an opportunity to participate. Williams was also warned to be certain that the group had German government support.[6] In December 1912 a further offer was made behind which, the Foreign Office decided, were Schröder and Tiarks, who wanted control of the railway as part of their larger scheme.[7] The offer was not pursued and nothing practical ever came of Schröder and Tiarks's plans.

As money was apparently not available from any other source, Williams began to consider more carefully the possibility of securing

[3] Williams to Grey, 31 May 1912 and minutes thereon, FO 367/290/15211, also various correspondence in FO 367/289/1841.

[4] Memorandum by Grey, 30 May 1912, FO 367/290/15211. An interesting but not well-elucidated point arising at this time concerns a financial connection between Edmund Davis and the Société Générale. Davis had been told he could come in on some of the Belgian bank's business. Williams was not told precisely what business this was, but because of this and Davis's continued involvement with German interests in southern Africa, he was convinced he would have to be extremely careful about allowing any foreign capital at all into the railway (Williams to Grey, 17 June 1912, FO 367/290/15211). Detailed information about Davis's connections with German and Belgian finance is unfortunately not available.

[5] FO to Williams, 12 June 1912, FO 367/290/15211.

[6] Memorandum by Grey, 20 June 1912, FO 367/290/15211.

[7] Williams to Grey, 4 Dec. 1912 and minutes thereon, FO 367/290/15211.

German finance. On advice from Grey, who had discussed the matter with the German Ambassador, Prince Lichnowsky, Williams went to Berlin in March 1913, armed with a letter of introduction to Solf from Richard von Kuhlmann, Counsellor at the German Embassy in London.[8] Cordially received by Solf and Sir Edward Goschen, British Ambassador in Berlin, 1908–14, Williams met a number of officials from the Colonial and Foreign Ministries and directors of the Deutsche Bank. From discussions with Arthur von Gwinner, one of the bank directors, and Baron Wilhelm von Stumm, Political Director of the Foreign Ministry, Williams drew the conclusion that some mutually acceptable arrangements could be made. He told the Germans that he would have allowed them to participate in the railway sooner had he known they were interested and would wait until May for the Bank to put forward specific proposals. In the meantime he would put off anyone else who might approach him.[9] This was bravado on Williams's part as there is no evidence whatever of there being anyone else interested in financing the railway at this time.

Williams emphasized that no agreement would be acceptable if it involved the assumption of control over the railway by German financiers. He indicated his willingness to sell Tanks's entire interest in the line under certain conditions, provided that both the Portuguese and British governments approved the sale. On the other hand he said he was prepared to sell a partial interest without consulting the Portuguese at all, provided that Pauling was given the contract to complete the line.[10] It is most unlikely that Williams intended these suggestions seriously; he undoubtedly had no intention of giving up any of his interest in the line. He carried on negotiations with the Germans, most probably as a means of 'blackmailing' the Portuguese government who were suspicious—to say the least—of German intentions in Angola to co-operate with him and provide financial support for the railway.[11] The Deutsche Bank insisted on having German government support for their participation in the railway venture, which Solf managed to convince Chancellor Bethmann-Hollweg to guarantee in June 1913. A consortium headed by the

[8] Crowe to Williams, 21 Jan. 1913, and Williams to Grey, 1 Mar. 1913, FO 367/342/3217.

[9] Williams to Grey, 6 Mar. 1913, and 'Memorandum re Benguela Railway' enclosed in Williams to Langley, 29 Jan. 1913, FO 367/2440/3323.

[10] 'Memorandum of Talk with 'K[uhlmann] at Drumtochty' (then Williams's home in Scotland), 26 Aug. 1913, Williams Papers.

[11] Carnegie to Grey, No. 102, confidential, 23 June 1914, FO 371/2085/6995.

Deutsche Bank and the Société Générale was to be formed to finance construction. The Germans insisted on controlling the line, which Williams was not prepared to allow. Initially he had the support of the Portuguese government who refused to allow German capital into the railway company. Negotiations came to an end in April 1914. Williams then began negotiations directly with the Société Générale.[12]

From these talks emerged a plan for the formation of a new company to carry on construction, reaching the frontier within three years and Kambove two years later. Except in certain circumstances the Benguela Railway Company was to pay no interest in any form on the debentures to be issued by the new company until Katanga mineral traffic had been secured over the entire line for at least one year. Until then the debentures created a charge only on the portion of the line actually built by the construction company. The Belgian group financing the company, represented by the Société Générale, was to have a fair number of seats on both the Lisbon Board and the London Committee of the Benguela Railway Company. If they took up a two-year option to purchase 1,350,000 Benguela Railway Company shares for £1 each plus 5 per cent interest calculated from the time Tanks had first taken them, they were to have representation on the Board equal to Tanks's. As the Portuguese government had changed its mind about German participation, Williams now had to agree to allow the Deutsche Bank to participate in the formation of the construction company by the Société Générale, although the Deutsche Bank was not directly involved in the proposal.[13] These arrangements had not been completed when war broke out and put an end to them.

PROBLEMS ON THE EVE OF WAR

While negotiations with the Deutsche Bank were still in progress it appeared to many people that a German take-over of the Benguela Railway was inevitable. Chartered began to doubt their conviction that the line would never be completed and to look for a way out of the 1899 railway agreement. They were sure that if the Germans controlled the line they would insist on the rigid enforcement of the agreement so as to secure all Katanga traffic, of which Chartered had

[12] Willequet, pp. 293–4.

[13] Williams to E. Francqui, 21 July 1914, BR 21 and Williams to Langley, 14 Oct. 1914, FO 371/2085/6995.

been enjoying a virtual monopoly. Initially the company argued that the agreement could not apply to the link between the Benguela Railway and the Congo's rail system, as these were not in British territory, and because Chartered had not contributed in any way to the construction of the Benguela Railway which was in fact in opposition to their interests. The practical result of the severance to which they had so readily agreed in 1909 would be to cut off Rhodesia's railways from north–south traffic with the Congo, not the original intention of the 1899 agreement. The company wanted government backing in negotiations for the cancellation or restricted interpretation of the railway agreement as part of the larger issues being discussed in connection with the revision of the 'Secret' Treaty.

One of the more junior Colonial Office officials thought Chartered had a very good case, but Sir John Anderson, Permanent Under-Secretary, maintained a thread of continuity with the department's earlier position, commenting: 'the Lobito Bay chickens are now coming home to roost'. Chartered had no case at all he felt, as any railway that might ever have been built in violation of the agreement would have had to go through foreign territory. If the company could now argue successfully that the construction of a foreign railway linking with theirs was beyond the scope of the agreement, such an interpretation would have made the agreement worthless from the beginning. Anderson did think, however, that an abrogation of the relevant clause would be welcome, but as Chartered had made the agreement with Germany on their own, it was up to them to try their hand again. 'Dr. Jim [Jameson] might be as successful as Mr. Rhodes.' A suggestion which would undoubtedly have been most galling to Chartered was that the best approach would be for them to raise the question through Williams and 'his teutonic allies'.[14]

The Foreign Office agreed and, although Grey was willing to approach the German government, felt the initiative should come from Chartered.[15] This the company welcomed, but turned the question back by saying they were not prepared at the time to make any definite suggestions but were ready to give careful consideration to any ideas Grey might have.[16] Although by this time negotiations between Williams and the Deutsche Bank had come to an end and

[14] BSAC to CO, 30 Jan. 1914, and minutes thereon by H. N. Tait, Clerk, Dominions Division, and Sir John Anderson, CO 417/550.

[15] FO to CO, No. 6995, 25 Mar. 1914, CO 417/550.

[16] BSAC to CO, 5 May 1914, CO 417/552.

the matter was no longer as pressing as it had appeared, some discussion continued of the way in which the matter could best be raised without making the British appear as suppliants.[17] One possibility was to put the point in the context of Germany's desire to participate in a scheme then being considered by the Portuguese for the development of Angola, particularly in the part of the plan involving railway development. The decision was finally taken to 'let sleeping dogs lie' until the completion—still hypothetical—of the Benguela Railway. Germany would then be put in the position of plaintiff; if they insisted on the enforcement of the agreement, Chartered could request arbitration or propose to sever the line either by establishing a double-terminal system at Ruwe or by regulations forbidding through traffic to the east coast. If the Germans sought British government support the Foreign Office could simply say that the company's interpretation of the agreement differed from the Germans', that the Foreign Office did not feel called upon to express an opinion, and could only act when the interpretation of the agreement had been settled by arbitration.[18]

The dog was allowed to slumber as peacefully as possible until July 1918 when Chartered asked the Foreign Office to confirm that the 1899 agreement would be regarded as having been abrogated by the outbreak of war.[19] The Foreign Office was prepared to agree to this, but 'someone in the CO [was] being fussy', raising the question of whether the agreement was in fact a treaty which would be cancelled automatically by the war, or a private agreement which would not be at all affected.[20] After lengthy argument on various points and consideration of the report of the Treaty Revision Committee it was decided that the agreement was a treaty and that it was 'not desirable that it should be revived at the end of the war'.[21]

The scheme for general development in Angola referred to by the Foreign Office was embodied in a bill introduced into the Cortes by the Colonial Minister in May 1914.[22] It was a wide-ranging plan, and Williams tried to secure part of the £8,000,000 to be raised to carry it through for his railway. He put a proposal to the Portuguese

[17] Minutes on CO to FO, No. 16584, 14 May 1914, FO 371/2085/6995.

[18] FO minute, 17 July 1914, FO 371/2085/6995, and Minute on FO to CO, No. 21747, 25 May 1914, CO 417/548.

[19] BSAC to FO, 9 July 1918, FO 371/3375/121642.

[20] CO to FO, No. 36061, 18 Aug. 1918, FO 371/3375/121642.

[21] FO to BSAC, No. 121642, 2 Nov. 1918, FO 371/3375/121642.

[22] Carnegie to Grey, No. 88, 21 May 1914, FO 371/2185/6995.

government and soon told Jadot that both the Foreign and Colonial Ministers seemed certain that the Cortes would approve the scheme and that the railway company would be helped to complete the line.[23] The bill was in fact passed by a cabinet meeting but got no further.[24]

Williams's financial position on the eve of war was becoming rather serious, quite apart from the problems of securing finance for the rest of the line. He needed some £526,000 in cash to meet loan instalments and debenture interest payments about to fall due.[25] He tried to arrange for a Portuguese group to take up some Benguela Railway debentures to provide this money, but war broke out before this could be done.[26] More fruitful were his attempts to get some help from the Société Générale, despite the fact that there had been some disagreement between him and the bank regarding Union Minière financial policy. The Société Générale agreed to loan the Zambesia Exploring Company £120,000 against the security of 7,000 Union Minière shares which that company held.[27] When the war started, only half the loan had been completed.[28]

Williams also turned once again to the British government, suggesting to Grey that they should replace the Belgian and German governments who were to have backed the Deutsche Bank/Société Générale consortium. He did not want large amounts of cash, but only a guarantee of interest on £1,800,000 4 per cent debentures to cover a contract with Pauling to complete the line. In return Williams would spend £1,000,000 for material and supplies in Britain, thereby securing for the country, he claimed, the 'key to the whole of Central and South Africa as well as a harbour that could accommodate almost the entire British fleet'.[29] Some people at the Foreign Office, aware of the potential military importance of Katanga minerals for Britain, were sympathetic. They thought the railway would eventually be built and that it was better for British capital to remain in control. Sir Walter Langley, Assistant Under-Secretary, pointed out, however, that such arguments were not the kind that appealed to the Colonial Office whose view, since British possessions were involved,

[23] Williams to Jadot, 25 and 26 May 1914, SGB File.
[24] Ibid., 1 June 1914, ZE 7.
[25] Undated, unsigned memorandum on the financial position of the railway in BR 21.
[26] Williams to Robert Williams & Co., 9 June 1914, ZE 7.
[27] Agreement dated 19 June 1914 and SGB to ZEC, 24 June 1914, ZE 7.
[28] Williams to Carton de Wiart, 14 June 1916, ZE 7.
[29] Williams to Grey, 25 Aug. 1914, FO 371/2085/6995.

would have to prevail. Chartered continued to maintain that the Benguela Railway would work against the interests of the Rhodesian railways; the government could therefore not take the kind of action Williams requested.[30]

Williams claimed that without some financial assistance, Tanks would have to go into liquidation.[31] German interests were reported to be trying to gain control of the railway through the Portuguese government, neutral in the early stages of the war, in part by emphasizing the precariousness of Williams's financial position,[32] and Williams was afraid that the German government would try to convince the Portuguese to cancel the concession entirely. He was clearly trying to 'run for all it [was] worth' his contention that his difficulties were in part the Foreign Office's fault for having advised him to seek German capital. In the light of the careful way in which Grey had spoken to Williams on this point and the fact that there was no concrete evidence of any other finance becoming available, it is not surprising that nobody in the Foreign Office took this contention seriously.[33] Further reason not to give it much consideration was provided by Carnegie who reported that Col. Freire d'Andrade, the Foreign Minister, who was anti-German and opposed any German plans for obtaining footholds in Angola, had told him that there was no intention on the part of the government of cancelling the concession. At the same time Carnegie did provide some support for Williams by pointing out that although the war had brought an end to all Portuguese negotiations with the Germans regarding finance in Angola, some of the Portuguese finance houses were largely under German influence, and might have been trying to gain control of the railway.[34]

Williams pressed his case, suggesting as an alternative to the guarantee that the government loan him enough money to tide the railway company over the war period. The Colonial Office saw no reason to modify their opposition as they did not feel the Benguela Railway was a colonial interest. Langley summed up the position by saying: 'There could of course be no question of the Treasury finding money for a Railway through a Portuguese Colony to tap mines in a

[30] CO to FO, No. 33956, 12 Sept. 1914 and minutes thereon, FO 371/2085/6995.
[31] Memorandum by Langley, 7 Oct. 1914, FO 371/2085/6995.
[32] Williams to Langley, 7 Oct. 1914, FO 371/2085/6995.
[33] Ibid., 14 Oct. 1914 and minutes thereon, FO 371/2085/6995.
[34] Carnegie to Grey, No. 180, confidential, 1 Nov. 1914, FO 371/2085/6995.

Belgian Colony, and as the CO will not help Mr. Williams he will have to go elsewhere for his money which will, I fear, not be forthcoming, which for many reasons is unfortunate.'[35] The Colonial Office acting on behalf of Rhodesia and South Africa continued to oppose any official British support for the Benguela Railway even after the war, and some divergence of view between the Foreign and Colonial Offices remained in evidence. The fact that Williams's business interests could not be identified with vital British strategic or political interests in Africa was at the crux of his difficulties with regard to the railway, after it was completed as well as while it was being built.

Whether or not it would be possible to retain the concession was an important consideration for Williams, but more immediate and pressing was the need to find £122,000 to pay debenture interest due at the end of 1914. It was impossible for the Société Générale to provide any further advances against the amount they had agreed to loan to Zambesia Exploring Company[36] and no funds were available from any other source. Details of the debenture position at the end of 1914 were as follows: The Benguela Railway Company had issued a total of £2,500,000 5 per cent debentures, authorized on 6 September 1909. The first part of the issue was authorized by the Portuguese government in April 1910. Before the war broke out, Series B, C, and D had been issued in amounts of £210,000, £850,000, and £790,000 respectively, Tanks guaranteeing payment on all the debentures for twenty years. Of the total issued, Tanks itself held £78,000, Zambesia Exploring held £1,028,000. Tanks had outstanding £2,176,480 5 per cent first mortgage debentures of its own, of which Zambesia Exploring held £11,704; £42,250 5½ per cent second mortgage debentures, and £2,620 in fractional debenture certificates. These were secured by a specific first charge on Tanks's holdings in the Benguela Railway Company, Union Minière, consols, and the London, Brighton, and South Coast Railway Company, as well as by a first floating charge on all other Tanks assets except those in Rhodesia. The Zambesia Exploring Company had outstanding £230,630.5*s* 5½ per cent debentures secured by a specific charge on 100,000 Tanks shares and a floating charge on the company's other assets.

The problem was resolved by convincing those of the debenture holders who could attend a meeting in England to agree to extend the

[35] CO to FO, No. 46882, 10 Dec. 1914 and minutes thereon, FO 371/2085/6995.
[36] Williams to Langley, 12 Nov. 1914, FO 371/2085/6995.

date for redemption of the Zambesia Exploring Company's own debenture debt of £229,092 for two years or until six months after the cessation of hostilities, whichever was later. Debenture holders were to receive a 5 per cent cash bonus, and the redemption price was raised from £105 to £106. The Trustees for the debenture holders were given a further £200,000 of the Benguela Railway Company's 5 per cent debentures. Tanks called a similar meeting reaching agreement with those debenture holders available—not a majority as many lived in Belgium. Interest payments were postponed until six months after the war, with postponed interest carrying a further 5 per cent interest.[37]

During the war construction work on the Benguela Railway ceased entirely as it was impossible to purchase permanent way material and rolling stock, even if money had been available for them. Local traffic developed sufficiently to cover working expenditure, but not to cover the payment of debenture interest.

[37] ZEC and TCL Annual Reports for years ending 31 Dec. 1914 and 31 Dec. 1915. The CFK link to Lake Tanganyika was completed during the war with the help of a grant from the British government.

CHAPTER VI

Post-War Delays

AFTER the war supplies of permanent way material and rolling stock were once again available. The only obstacle to completing the Benguela Railway rapidly and easily to the Katanga frontier was the lack of money. Even the fact that the estimated cost of construction had more than doubled—from £1,703,000 in 1914 to £3,882,647 in 1919[1]—would not have made this an insurmountable obstacle if the Benguela Railway Company had been able to issue new debentures on attractive terms and convince potential investors that the line would in fact be connected to the mines and would carry sufficient traffic to begin showing a profit within a reasonable period of time. This required close co-operation between Williams and the Portuguese government, certainty that the Belgians would honour their commitment to build the section of the line in the Congo, and the expansion of Union Minière's output. The first of these conditions was extremely difficult to achieve, the second continued to be questioned despite official assurances, while the third was fulfilled fairly quickly.

UNION MINIÈRE EXPANDS

Relations between Williams and the Portuguese government at the end of the war were rather strained. In June 1913 the Cortes had passed a law that made it possible for the Benguela Railway Company to pay for materials and construction as well as debenture servicing costs with debentures, without necessarily transferring construction and operation of the line to debenture holders. A decree issued in June 1918 reversed the 1913 debenture law, ostensibly to prevent all colonial railways from being mortgaged, but in fact to make it impossible for the Benguela Railway Company to put new debentures on the market. Holders of the existing debt could hardly be expected to agree to a new issue secured on very different terms. The only

[1] Pauling & Co. to BRC, undated copy in G. C. Hutchinson, Williams's son-in-law and assistant Managing Director of Tanks and ZEC, to FO, 15 Dec. 1919, FO 371/4127/108013.

response to Williams's urgent requests that the decree be rescinded was the establishment of a Commission of Enquiry that recommended referring the matter to the Cortes; effectively that nothing at all be done. Approaches through the Foreign Office produced no more practical results[2] and neither the British nor the Portuguese delegates to the Paris Peace Conference considered the question sufficiently important to be discussed there.[3]

In August 1919 William Seeds, British chargé d'affaires in Lisbon, put Williams's case once more to Freire d'Andrade. The government was prepared to draft a law repealing the decree but giving them first option on any debentures issued. The railway company could not object to this, but did object very strongly to the fact that no time limit was to be set for the exercise of the option. The government could delay a decision indefinitely, during which time the debentures could not be sold elsewhere; the line would become uneconomical to operate and would ultimately be taken over by existing debenture holders.[4] Freire d'Andrade did nothing to press the draft law through the Cortes, although he assured Seeds that the government were aware of the railway's hardships and would do their best to pass the law during the next parliamentary sessions.[5] What Freire d'Andrade really wanted to do was to force Williams to agree to the formation of a new company to carry out the remaining construction work. On the assumption that the Belgian government wanted the line to be completed, he wanted Belgian finance in the new company and was quite prepared to allow Belgian interests to take over the line entirely.[6]

There was no doubt that the Belgians did want the Benguela Railway completed as they had plans for considerable expansion of output. Following the war, American copper producers who exercised virtually complete control over production in Canada and Latin America as well as in the United States, tried to keep copper prices high in the face of a number of circumstances tending very strongly to

[2] BRC to Sir W. G. Tyrrell, Assistant Under-Secretary FO, 25 July 1919, FO 371/4129/108013.

[3] BRC to Tyrrell, 30 July 1919, and Balfour to Curzon, 13 Aug. 1910, FO 371/4127/108013.

[4] Memorandum, 20 Aug. 1919, copy in Seeds to Curzon, No. 110, 20 Aug. 1919, FO 371/4127/108013.

[5] Seeds to Curzon, No. 120, 5 Sept. 1919, FO 371/4127/108013.

[6] R. Leghait, Belgian Ambassador in Lisbon, to P. Hymans, Foreign Minister, confidential, 13 July 1919, and F. Cattier to Hymans, 23 Dec. 1919, MAE, Af 1–9.

depress them. Copper stocks had been accumulated to meet actual and anticipated military needs; when hostilities ceased it was expected that the sudden drop in consumption would be offset by civilian demand expanding from wartime restriction. There was, however, little chance of Europe's absorbing American stockpiles or new output in the immediate future. European governments held about 240,000 tons of virgin copper and about 400,000 in brass and other scrap, to say nothing of an estimated 1,000,000 tons littering and recoverable from the battlefields.

In an attempt to reduce the amount of copper coming on to the world market, major United States producers, representing between 89 and 90 per cent of the country's output, in December 1918 formed the Copper Exporters Association. To qualify for exemption from the anti-trust laws the Association's stated purpose was to act as export broker for members, distributing what orders were received in proportion to each one's production. Despite vigorous denials, there is no doubt that the members of the Association tried to keep prices at a profitable level by cutting down production and buying up government stocks. They had little competition from other countries. Germany, Japan, and Russia, who before the war had exported appreciable amounts of copper, had become importers. Other areas such as Spain were not in a position to increase their plant capacity significantly. Katanga was the only major threat to American control of world prices, particularly as all of Union Minière's output went on the world market, none being used within the Congo itself.[7]

By the end of 1918 Union Minière's copper output had reached a total of 105,528 metric tons. About 85,000 tons of this had been produced during the war and had all been bought by the British government. The inflated wartime price of copper had made it possible for the company to install four additional water-jacket smelting furnaces and to pay a dividend of 150 francs per share for the five years up to and including 1918. (Because of the German occupation of Belgium it had been impossible to pay dividends during the war,

[7] Average copper prices in New York for the period 1918–27 were: 1918—$0·247 per pound; 1919—0·186, 1920—0·184, 1921—0·129, 1922—0·135, 1923—0·147, 1924—0·131, 1925—0·142, 1926—0·140, 1927—0·131. US Department of Commerce, Bureau of Mines, *Mineral Resources of the United States, 1927, Part 1, Metals* (Washington, D.C., 1928), p. 729. Alex Skelton, 'Copper' in W. Y. Elliott *et al.*, *International Control in the Non-Ferrous Metals* (Macmillan, New York, 1937), pp. 415–27; Orris C. Herfindahl, *Copper Costs and Prices, 1870–1957* (Johns Hopkins University Press, Baltimore, Md., 1959), pp. 92–6.

much to Williams's distress.) High production levels had been possible only because of the 'scandalously' rich ores that were worked. These sometimes contained as much as 25 per cent copper, although the average content was about 15 per cent, compared with the less than 2 per cent found generally in American ores and the 3–4 per cent of the Copperbelt. These ores could go directly to the smelters without preliminary treatment. By 1918 rich reserves had been seriously depleted and lower-grade ores had to be concentrated before smelting. This could be done mechanically by crushing and sorting the ores, but the job could be done more efficiently with the recently developed flotation process in which crushed ores are put into a vat of oil which is then whipped up causing the lighter copper-bearing particles to rise to the surface on the froth. Some concentrating equipment was essential if production was to continue at all, but much more than that was needed for any major expansion.

Union Minière decided to increase their output substantially when American producers were decreasing theirs in order to take advantage of the artificially maintained prices, and looking forward to the time when prices would rise again. The most effective means for carrying out this expansion was to use the electro-chemical leaching process in which ores are dissolved in sulphuric acid, the copper then being recovered by electrolysis. This technique, particularly suitable for treating sulphide ores, had been perfected during the war in Chile and the United States. If the power needed for a leaching plant could be provided cheaply enough, this method of refining had several advantages for Union Minière. It would enable them to use lower-grade ores to produce higher-quality copper than they could with their smelters, would eliminate the need to send all their copper to the United States or Germany for further treatment, and would substantially lessen their independence on supplies of coal and coke from Wankie. Thermal generators had been fed with coal from the Luena coalfields in Katanga, which did not produce good metallurgical coke. Sufficiently cheap power, however, required the harnessing of the Lualaba or Lufira Rivers. Because of the high cost of hydro-electric installations, the company decided as a first stage in their expansion programme to go ahead with the construction of concentrating equipment but to install only a pilot leaching plant. This limited scheme was financed by the creation of 25,000 Union Minière capital shares and a further 25,000 dividend shares which were turned over to the CSK. The capital shares had a par value of 100 francs and were issued at 800. Tanks took up 9,800 shares at this price, receiving 10,000 dividend shares free. The Zambesia Exploring Company took up 1,100 capital shares, raising their Union Minière holding to 12,100 shares.[8] The wider expansion plans were left in abeyance.

Apart from any financial obstacles to this expansion, there was little point in increasing output if the company could not be assured of having adequate facilities to transport what it produced to the coast. The CFK/Rhodesian Railways link to Beira, supplemented to a small extent by the rail/water route to Dar-es-Salaam via Lake Tanganyika, had proved adequate up to this time, but could not be expected to cope with very much more. The BCK line was still only projected and no definite steps had yet been taken to begin its construction. The Benguela Railway was of major importance. In March 1919 Louis Franck, who had become Belgian Colonial Minister in 1918, confirmed the government's intention to honour its obligations under the 1908 'Pooling Agreement', to link the Benguela Railway to the CFK,[9] but was not very much in favour of any direct Belgian participation in the Benguela Railway itself. He felt the Belgians should not 'put themselves on Williams' back' by trying to supplant him, but that any money Belgium might have for railway construction should be devoted to the BCK. The Foreign Ministry thought, however, that they should co-operate closely with the Portuguese to ensure the completion of the line, not only because it would advance Katanga's development but also because it would help free Katanga from South African influences and create a favourable political climate for raising some questions with Portugal, notably the cession of some Angolan territory on the left bank of the Congo which was needed for the improvement of port and harbour facilities at Matadi.[10]

The Foreign Ministry view prevailed, at least enough for the government to ask the Société Générale and its affiliate the Banque d'Outremer to provide the necessary financial backing for Freire d'Andrade's scheme. Jadot was quite prepared to do this but wanted to reorganize the undertaking in co-operation with both the Portuguese and Williams. There were, he stressed, considerable financial problems posed both by the railway company's existing burden of debt and by the large amount of capital needed to construct the

[8] ZEC, TCL, and UMHK Annual Reports for 1919 and 1920.

[9] Memorandum 'Interview . . . Franck . . . Williams', 12 Mar. 1919, copy left at FO by Williams, 4 Feb. 1920, FO 371/4127/108013.

[10] 'Note pour M. le Ministre', 24 Dec. 1919, MAE, AF 1—9.

remaining section.[11] Jadot also urged Williams to accept a proposal put forward by the Banco Nacional Ultramarino (BNU) in which the Société Générale would participate and which had Freire d'Andrade's support. A consortium of banks would take up a new issue of Benguela Railway Company shares; the BNU would then issue an equivalent amount of 5 per cent debentures, 3 per cent guaranteed by the bank, 2 per cent by the Portuguese government.[12] This proposal, as far as Jadot could see, indicated a new desire on the part of the Portuguese government to assist the railway. Williams maintained that no such desire was in evidence, but to the contrary the Portuguese were still being obstructive. The BNU's plan would greatly depreciate the value of Tanks's holding in the railway and was entirely unacceptable. If Jadot really wanted to help, Williams said, he should support a £4,000,000 debenture issue by the railway company, guaranteed by Tanks who would then be able to retain 'control'. 'Monsieur Jadot did not relish this scheme.'[13]

Both men were, from their own points of view, right. Jadot, who wanted the railway completed as quickly as possible, thought the scheme a good one because it clearly had official support and seemed as certain as anything involving the frequently changing Portuguese government could be, to be carried out with a minimum of obstruction and a maximum of co-operation. He was of course not so concerned about the extent of Tanks's interest in the company as Williams was, although he did not want Williams pushed out entirely. Williams on the other hand was not solely concerned with seeing the Benguela Railway finished, but also wanted to retain his company's majority shareholding in it. The question of control to which he so often alluded had little real importance as Portuguese nationals made up a majority of the Board of Directors and could override Williams on policy questions at any time. What obviously did matter was the 90 per cent share of the substantial profits he still confidently expected the railway to earn. When Firmin van Bree, a close associate of Jadot and also a Union Minière director, put forward an alternative proposal involving French and American capital as well as Belgian and Portuguese, Williams replied that this would be unacceptable

[11] Jadot to Hymans, 7 and 21 Jan. 1920 and 'Note pour M. le Ministre à Lisbon', 27 Feb. 1920, MAE, AF 1–9.

[12] Lane to Williams, 19 Sept. 1919, copy in Hutchinson to FO, 3 Oct. 1919, FO 371/4127/108013, and Leghait to Hymans, 19 Dec. 1919, MAE, AF 9.

[13] Memorandum 'Meeting . . . Société Générale . . .', 26 Jan. 1920, copy left at FO by Williams, 4 Feb. 1920, FO 371/4127/108013.

because of his agreement not to bring foreign capital into the enterprise without permission of the Portuguese government.[14] This was the same argument he had used in connection with German capital before the war. He failed to explain why he was sure the Portuguese would reject the proposal, but added that in any case he and his British colleagues would manage to finance completion without outside help. Sir Cecil Budd's British Metals Corporation had enabled Tanks to increase its share capital by £1,000,000, and both Budd and Williams were quite sure they would get support from the British government and would have no difficulty in raising the money needed.[15]

Tentative agreement on the repeal of the debenture decree was reached with the Colonial Minister in March 1920 but due to one of the five changes of government that took place that year, it was only in August that the Cortes passed a law repealing the decree. To meet British government wishes, the Portuguese government's option to take up debentures had to be exercised within fifteen days of the Colonial Minister's being informed of the issue price. The railway had to be completed to kilometre 620 by the end of June 1922, to kilometre 720 within the next twelve months, and to the frontier by the end of June 1925.[16] Williams was then able to have shipped to Angola 100 kilometres of permanent way material, the money for which he had arranged to borrow from the London Joint City and Midland Bank.[17] This, when laid, left the railway more than 500 kilometres from the frontier. A relatively small bank loan was not the answer to Williams's financial difficulties. He still had to find about £4,000,000 to complete the line. The railway company could now issue debentures but it remained to be seen whether people could be found to purchase them.

SMUTS SAYS 'NO'

In May 1921 Williams was once again knocking at the Foreign Office door, arguing that access to Katanga copper was of the utmost

[14] Memorandum 'Meeting . . . Société Générale . . .', 26 Jan. 1920, copy left at FO by Williams, 4 Feb. 1920, FO 371/4127/108013.

[15] 'Resumé d'un entretien de M Jadot avec MM Robert Williams et Sir Cecil Budd le 26 janvier 1920', MAE, AF 1—9.

[16] Carnegie to Curzon, No. 103, 9 Apr. 1920, D. Keeling, chargé d'affaires in Lisbon, to Curzon, No. 231, 20 Aug. 1920, 'Cable from Lisbon', 19 May 1920, copy in Hutchinson to Tyrrell, 27 May 1920, FO 371/4127/108013.

[17] Hutchinson to Tyrrell, 16 July 1920, enclosing copy of a cable from Lisbon, 13 July 1920, FO 371/4127/108013.

strategic importance for Britain whose empire had no proved source of the metal. In order to ensure an adequate supply independent of America (presumably including Canada), the Benguela Railway had to be completed and attempts by Belgian, American, or German interests to take it over resisted. He added the rather extraordinary contention that the railway also had great strategic importance for the Sudan and Egypt if a future war should close the Mediterranean and the Suez Canal to British shipping. As the Belgian government had guaranteed to finance the Katanga section of the line to Elisabethville, Williams wanted the British to do the same for the rest of the Angola section, particularly as they had just guaranteed a debenture issue for a line from Beira to the Zambezi River through Moçambique. The two cases were of course not really comparable as the Shire Highlands Railway to which Williams was referring would provide an important access route for the British territory of Nyasaland, but whether Williams was aware of this difference or not he wanted a government guarantee for a debenture issue of £4,000,000, £2,300,000 of which would be spent on purchasing materials in Britain.[18]

The Board of Trade agreed that Katanga's resources were of great importance to British industry and that British interests should share as much as possible in their development, but they were not important enough to justify going beyond the 'usual diplomatic support accorded by His Majesty's Government in such cases'.[19] The Colonial Office was simply 'not aware of any special grounds' for supporting Williams's request.[20] Williams was very annoyed by this as the government would only have been required to pay out any money in the highly unlikely event of railway earnings being insufficient to cover debenture interest. This he considered a trifling liability in comparison to the enormous benefits which would accrue to British industry from the railway and Katanga's development. If, however, the government decision was final, Williams said he might be forced to seek American capital.[21]

Williams seems to have been either unwilling to admit or blissfully unaware of the many flaws in his argument. Even if one accepted

[18] Memorandum: 'Points for Foreign Office Interview', 18 May 1921, with minute by N. Bland to Crowe, FO 371/5515/3623. Bland was private secretary to Crowe who was at this time Permanent Under-Secretary.

[19] BT to FO, P.T.E. 1783, 25 July 1921, FO 371/5515/3623.

[20] CO to FO No. 35583, 23 July 1921, FO 371/5515/3623.

[21] Williams to FO, 9 Aug. 1921, FO 371/5515/3623.

the premisses—by no means unquestionable—that an assured supply of copper independent of America and an alternate route to the Sudan were vital to Britain, it did not necessarily follow that the Benguela Railway had to be built. The Belgians could exercise absolute control over the copper supply; if they remained willing to sell their copper to Britain the existing lines provided at least some assured transport. Further transport facilities under direct British control were available in Tanganyika.[22] These existing lines as well as the Kenya and Uganda rail systems provided access to the Sudan and Egypt in some ways much better than the Benguela Railway as they did not depend on transit through the Congo. There was no British national interest which could be served by the Benguela Railway. Apart from the political advantage the Portuguese government could expect to derive from the economic benefits that would inevitably accrue to them when the line was completed and which they had already begun to receive through taxation and local development fostered by the already completed section, the Benguela Railway's *raison d'être* was still the economical transportation of Katanga minerals. For that to be achieved close co-operation between the railway company, the Belgians, and the Portuguese was vital.

It is also difficult to understand why Williams thought the British government would be moved by his threat to seek American capital, especially as he had often indicated his extreme reluctance to do so. He may have thought that as the use of American money would undoubtedly have involved the purchase of materials in the United States, the British government would support him to secure orders for British firms. If that was the case he would have had to provide convincing proof that American finance would in fact be available to him. Williams did send the Foreign Office a copy of a letter from John Hays Hammond, the American mining expert whom Williams had known earlier in South Africa, suggesting that Hammond's

[22] In 1921 the Belgian and British governments formalized an agreement reached in principle in 1917 for the free passage of goods between Dar-es-Salaam and the Congo and Ruanda-Urundi. Perpetual concessions at Dar and Kigoma were also granted, the former operated by the Agence commerciale Belge de l'Est Africain (Belbase) which acted as shipping agent. In 1922 4,000 tons of freight to and from Belgian territory passed through Tanganyika. By 1930 traffic had risen to 65,000 tons, then dropping off considerably until the Second World War. Most of this was general commercial traffic, but the route could have been further developed to take larger amounts of copper traffic than it in fact did. André Huybrechts, 'Les Voies d'accès et d'évacuation du Congo', *Cahiers Économiques et Sociaux*, vii (1969), 3.

American group should be brought into the Katanga copper industry. As the letter had been written in 1906 and Williams had refused the offer, this could hardly have had much of an impact. In the intervening years Hammond had become connected with a number of companies whose interests Williams regarded as inimical to his own. One approach was in fact made to Speyer & Co. in New York, but they refused to take any interest in Williams's scheme as they considered the 'American public . . . not yet sufficiently educated up to subscribe to such an issue'.[23] American financiers were not the solution to the problem.

Williams's next move was to put his proposals for a debenture guarantee to Sir Alfred Mond, Chairman of the Cabinet Committee on Unemployment Relief, who was willing to support Williams's request.[24] Before Mond's Committee could consider the matter, Parliament passed the Trade Facilities Act authorizing Treasury guarantees of debentures issued by British or Foreign companies ordering materials in Britain. The Treasury could only give these on the advice of a committee appointed by Parliament to advise on the Act's implementation.[25] The Advisory Committee's Chairman was Sir Robert (later Baron) Kindersley, K.B.E., G.B.E., then a Director of the Bank of England and President of the National Savings Committee. Other members were Sir William (later Baron) Plender, K.B., G.B.E., senior partner in the accounting firm of Deloitte, Plender, Griffiths & Co., who had served on a number of government inquiry committees, and Lt.-Col. (later Sir) George Schuster, C.B.E., M.C., Chief Assistant to the organizer of the League of Nations International Credit Scheme. Williams immediately applied to the Advisory Committee for approval of his plans.[26]

The Committee were rather dismayed by the proposal to spend only half the £4,000,000 issue in Britain, and raised some obvious questions about the Portuguese right to purchase the line, which they felt might endanger security for debenture holders, the Belgian intention to complete the line, and the adequacy of traffic from Union

[23] Précis sheet, 'Benguela Railway Negotiations', undated, Files of Schröder, Wagg & Co., London.

[24] Williams to Mond, 30 Sept. 1921, TFN, and Williams to Crowe, 29 Sept. 1921, FO 371/5515/3623.

[25] 11 and 12 Geo 5, Ch. 65. The Act was extended in 1922 (13 Geo 5, Ch. 65), 1924 (14 and 15 Geo 5, Ch. 8), 1925 (15 Geo 5, Ch. 13), and 1926 (16 Geo 5, Ch. 3).

[26] Williams to Kindersley, 11 Nov. 1921, TFN.

Minière.[27] Williams tried to reassure the Committee on these points and after receiving a letter raising some further points of detail, cabled—through the Foreign Office—to Lane in Lisbon that he had provisionally arranged the finance needed to complete construction.[28] Once again his optimism was premature.

By this time, Williams's relations with his 'Belgian Banker Colleagues' had deteriorated to the point where he expressed 'grave doubts about [their] loyalty' and accused Émile Francqui, who had become a Director of the Banque d'Outremer and was active in Forminière and the Companhia de Diamantes de Angola, of having instigated the 1918 decree.[29] He may well have been right. There was no reason why the Belgians should have felt particularly loyal to Williams when what they saw as their vital interests were at stake. He had refused to accept Belgian proposals for completing the line but had so far proved incapable of managing on his own. The need for additional transport facilities was becoming increasingly urgent, and Williams would have to look after himself. His Belgian associates were regarded as among the 'trickiest' and most powerful in the world; when he came up against men like Francqui and Jadot, he was, according to Sir Gerald Grahame, British Ambassador in Brussels, 'hopelessly outclassed'.[30]

Union Minière was once again considering its wider expansion plans. Before a final decision was reached, there was a further crisis in the world copper market. With the post-war 're-stocking boom' at an end, the demand for copper fell off so drastically that American and Canadian producers agreed to impose even greater restrictions on their output. Union Minière decided not to go along with this and to continue taking advantage of the artificially sustained prices. With the support of the Colonial Minister and the continued financial backing of the Société Générale, the company set itself the goal of doubling its output. In September 1921 an agreement was signed with the CSK calling for the creation of 111,000 Union Minière capital shares to be given along with 6,000,000 francs in cash to the CSK, which covered

[27] TFAAC to TCL and BRC, 2 Dec. 1921, TFN.

[28] Williams to Kindersley, 12 Dec. 1921, and TFAAC to TCL, 16 Dec. 1921, TFN and Williams to Lane, telegram, 17 Dec. 1921 in Williams to Bland, 19 Dec. 1921, FO 371/5515/3623.

[29] Williams to Tyrrell, private and confidential, 4 Apr. 1921, FO 371/5515/3623.

[30] E. Phipps, Counsellor at the British Embassy in Brussels, to Lindsay, Assistant Under-Secretary, FO, private and confidential 24 Dec. 1921, FO 371/6973/12607.

the increase of the company's capital by one-third as envisaged in the original grant from the CSK for 1936. In return Union Minière's concession was extended to 11 March 1990. Because of the state of the market, the new shares, also of 100 francs par value, were issued at only 400 francs. The agreement with the CSK also authorized the company to issue up to 150,000,000 francs in debentures or privileged shares. At the outbreak of war, Union Minière's debenture debt had been 15,000,000 francs at $4\frac{1}{2}$ per cent. In 1919 a further 5,000,000 francs was added to this and subsequently a new issue of 40,000,000 francs at 7 per cent was made. Instead of further increasing this debenture debt the company issued 300,000 privileged shares of 500 francs par value. Tanks was not able to take up any of these, but by the end of 1923 held 103,048 Union Minière capital shares, Zambesia Exploring holding another 16,235. Williams's companies then held about 16 per cent of Union Minière's nominal capital, compared with over 45 per cent they had held three years previously. Williams had not been able to maintain his position in the Belgian company because of the heavy commitment of his financial resources to the Benguela Railway.[31]

Not surprisingly then when Williams in a rather peremptory and condescending manner asked Grahame to tell Franck about the provisional arrangements for finance and 'get' Jadot to write a letter guaranteeing to provide the money for Union Minière's expansion programme, Grahame was more than a little reluctant to comply. As Williams had always dealt directly with these men, if Grahame suddenly appeared as an intermediary, the inference would be drawn that he was acting on instructions from Whitehall, or at least with Foreign Office approval[32]—undoubtedly the precise impression Williams wished to create. Grahame was authorized to deliver Williams's messages verbally, stressing that he was not expressing the official views of the British government. As predicted Franck was rather surprised that Grahame was involved at all, and Williams gained nothing.[33]

[31] ZEC, TCL, and UMHK Annual Reports for 1919–23; *Union Minière du Haut-Katanga . . .*, pp. 128–33.

[32] Williams to Grahame, strictly confidential, 19 Dec. 1921, copy in Williams to Bland, confidential, 22 Dec. 1921, FO 371/5515/3623, and Grahame to Williams, 21 Dec. 1912, copy in Grahame to Curzon, No. 1042, 21 Dec. 1921, FO 371/6973/12607. Grahame's dispatch also enclosed a copy of Williams's letter suitably edited to make its tone less officious than that of the original.

[33] Grahame to Curzon, No. 1054, 24 Dec. 1912, FO 371/6973/12607.

Grahame urged the Foreign Office to be very circumspect about giving Williams any further help, and in responding to any request for their views from the Trade Facilities Act Advisory Committee. He was not entirely sure of how Belgian financiers viewed the Benguela Railway at this stage, particularly as no definite plans had been made for building the BCK. Pressure to do this was, however, growing and consideration should, Grahame felt, be given to part of a recent speech by the Congo's Governor-General: 'Je dis avec vous à la Belgique: Plus un instant à perdre. Il faut qu'une ligne belge du Bas-Congo au Katanga soit ouverte et nous desserve *au plus tard* le jour même ou s'ouvrira la ligne de Lobito Bay.'[34] A copy of the letter conveying Grahame's advice was sent to the Advisory Committee with a specific request that its contents should not be disclosed to Williams.[35]

The Committee and Williams continued discussions to work out details of the proposed debenture issue and the amount of money to be spent in Britain. The Committee were officially competent only to consider the commercial aspect of any request put to them, but were not unaware of several political considerations involved. They indicated some fear that the political climate in Angola as well as in Portugal itself might bring about a situation in which the British government would feel compelled to intervene to protect their guarantee. More immediate and important was the disturbance felt by British interests connected with Rhodesia when rumours began to circulate in the City that the Committee were about to recommend the guarantee. Although the Committee did not think all the objections raised were justified, they did feel that Rhodesian attitudes should be considered and wanted to ascertain the government's position before proceding further.[36] Williams, unaware as he often seems to have been of the political aspects of what he did, could hardly have been very surprised by this turn of events. At the same time it is quite understandable that it put him in an embarrassing and difficult position.[37]

Rhodesia Railways admitted that they could not object to the completion of the Benguela Railway, but they could and did most

[34] Phipps to Lindsay, private and confidential, 24 Dec. 1921, FO 371/6973/12607.

[35] G. Villiers, Assistant Secretary, Western Department, to TFAAC, very confidential, 5 Jan. 1922, FO 371/6973/12607.

[36] TFAAC to Treasury, confidential, 16 Feb. 1922, T 160/547(3)/ 4273.

[37] Williams to TFAAC, 27 Feb. 1922, TFN.

vociferously object to the British government's accelerating the diversion of a large amount of coal and copper traffic then being carried over their lines. This traffic accounted for 23 per cent of the railways' revenue and made it possible for them to provide the Rhodesias with transport facilities they could not otherwise afford. Agricultural products, then becoming an increasingly important part of Southern Rhodesia's exports, would be adversely affected by the higher rates that would have to be charged for carrying them. The diversion of traffic to the Benguela Railway would also greatly reduce the Rhodesian lines' purchasing power—a substantial portion of which was devoted to buying stores and rolling stock in Britain—and jeopardize the annual payment of £734,000 in debenture interest secured on them. If the government gave the guarantee, they would then have a direct interest in creating this diversion.[38]

The Union of South Africa, especially its Prime Minister, General Smuts, also opposed the guarantee, in part because the Benguela Railway would enable Katanga merchants to import goods directly from Europe rather than from Johannesburg as they had been doing for many years,[39] but primarily because Smuts hoped to make Southern Rhodesia part of the Union, amalgamating Rhodesia's and the Union's railways. This, Smuts argued, would enable the Union to capture a large part of Katanga's traffic which would in turn help to establish British influence in the 'most progressive part [of the] Congo Free State [*sic*].' If Belgian or American finance were in fact available, the Benguela line would be built in any case, but the longer completion could be delayed, the better it would be for British investment in South Africa and Rhodesia.[40] Sir James Masterton-Smith, Permanent Under-Secretary at the Colonial Office, felt it their 'clear duty to oppose' the guarantee.[41] The Treasury was told that the objections raised by Smuts and Rhodesia Railways constituted serious obstacles to the guarantee and that the Colonial Office were 'strongly of opinion that in the circumstances a guarantee should not be given'.[42]

[38] Rhodesia Railway Memorandum, 21 Feb. 1922, FO 551/139.

[39] Opposition to the Benguela Railway had been expressed in the Cape, by the Chamber of Commerce, as early as 1906. (Cf. minutes on Williams to Churchill, 4 May 1906, CO 417/429.)

[40] Edgar Walton, South Africa's High Commissioner in London, to Winston Churchill, then Colonial Secretary, 20 Feb. 1922, CO 551/153.

[41] CO minutes, 17 Feb. 1922, CO 551/157.

[42] CO to Treasury, No. 8412, 3 Mar. 1922, T 160/547(3)/4273.

The Foreign Office took a rather different line, professing an inability to understand Smuts's objections, particularly as British interests would be very well served by the expenditure of about £3,000,000 in Britain. The more the Katanga mines were developed, the better it would be for Rhodesia Railways. Virtually all the food and labour for Katanga came from Rhodesia, they added, which was particularly important for Northern Rhodesia, whose farmers relied heavily on Katanga markets. There was no question of Belgian competition with the Benguela Railway as far as the Foreign Office could see because of the terms of the 'Pooling Agreement'. Possible Portuguese interference with the railway was thought unlikely in Angola, where there was no sizeable white population and no 'Bolshevik municipalities'. Should the need ever arise, representations for the removal of harassing regulations or intrusions would be made.[43] Neither this argument nor Williams's continued pressure changed the Colonial Office position. They felt they had to be guided in the matter by Smuts, and that it was out of the question for the British government to encourage British investment in a foreign country, the effect of which would be to depreciate the value of existing investment in British territory.[44]

British government assistance for the Benguela Railway was given a broader dimension by Libert Oury, a naturalized British subject born in Belgium, a Director of the Moçambique Company and Manager of its London Committee. In a confidential memorandum he maintained that Belgian colonial circles, the Belgian press, and the Belgian public were 'deeply agitated' by the rumours then circulating to the effect that the British government had decided to lend its support to the Benguela Railway. This was seen as a purely political act designed to retard or prevent the construction of the Congo's own national line, the BCK. Those Belgians who objected to the guarantee dismissed the 'Pooling Agreement' as having been reached under circumstances very different from those prevailing in the post-war period, and of little practical importance. If Williams could finance the completion of the Benguela Railway without government help, the Belgians would not object and would build

[43] Minute by Villiers, 27 Feb. 1922, FO 371/8237/265, and FO to Treasury, 3 Mar. 1922, T 160/547(3)/4237.

[44] Minute on Treasury to CO, 4237, 21 Feb. 1922, CO 551/156. Walton suggested that Williams go to Cape Town to discuss the matter directly with Smuts, but Williams claimed such an action would be misinterpreted in Belgium and Portugal.

their section of the line. Oury felt that in view of the 'grave misunderstanding' which might arise, very careful consideration should be given to the proposed guarantee.[45]

Oury, very much involved in the development of the port of Beira, was hardly a disinterested observer, but much of what he said was well founded. Anglo–Belgian relations at this time were somewhat strained on several counts,[46] and Williams was in a very unenviable position. The 'Pooling Agreement' on which he placed so much emphasis could be modified, and if the completion of the Benguela Railway could be delayed long enough, the link to the mines would not necessarily have to be built, and if it were, the BCK could be constructed rapidly and the pooling provisions of the 1908 agreement abrogated. If the BCK were completed without delay, there would be little reason for the Belgians to help Williams, particularly in view of his own unco-operative attitude earlier. The conviction held by Leopold and Williams in 1908 that the BCK, Benguela Railway, and the railway to Beira would all be needed for the full development of Katanga's wealth was being very seriously questioned. Arguments were advanced to show that the BCK and, less convincingly, the rail/river route provided by the CFL, would be able to carry the additional traffic resulting from Union Minière's expansion as cheaply or more cheaply than the Benguela Railway.[47]

The Foreign Office sent Grahame a copy of Oury's memorandum, asking what importance should be attached to it. From outside observation and from what had been published in the press by those who objected to the guarantee, Grahame said he would have been inclined to think the opposition somewhat artificial, being raised by a 'noisy group of colonials'. On further investigation, however, he found Oury's view of the situation very well justified, representing accurately the view held in 'responsible quarters'. The Belgian government did not harbour the suspicion that His Majesty's Government were trying to control communication between the Congo and the rest of the world, but the Belgian public were very much inclined to believe just that and could easily be convinced it was the case. Williams's references to the strategic importance of the

[45] Confidential memorandum enclosed in Oury to W. C. Bottomley, Assistant Secretary, CO East African Department, 4 Mar. 1922, CO 551/155.

[46] See Sally J. Marks, 'British Policy towards Belgium from the Armistice to the Fall of Lloyd George' (University of London, Ph.D. Thesis, 1968).

[47] Many articles appeared in the press on this subject, and organizations with colonial interests discussed it at length. Some ideas were not very realistic.

Benguela Railway for the Sudan 'got the Belgian goat'. Oury was also correct, Grahame thought, in saying that the Belgians would build their section of the line if they had to, but only with very bad grace. The Belgian authorities were now convinced that within four or five years the BCK would provide cheaper transportation than the Benguela Railway, and that the Benguela line would never pay and could not, therefore, attract American finance. Considerable difficulty, political and financial, could be expected in Belgium if the link had to be built. The Belgians would do so completely sceptical of the profit they might derive either for the company that built it or for Belgian colonial interests generally. There was no surprise at Smuts's opposition; his attitude was much like the Belgians' own.[48]

After further discussion with Williams and the Foreign and Colonial Offices, the Advisory Committee reported to the Treasury that on purely commercial grounds they were 'not unprepared to regard [Williams's] proposition favourably'.[49] The Treasury added its own support to this, as they were interested in 800,000 man-weeks of work for British shops which the scheme would provide.[50] The Colonial Office, when pressed further, admitted that Northern Rhodesia would probably benefit from the completed Benguela Railway and that Williams presented a good case. None the less, their position could not change unless Smuts removed his 'veto', and it was up to Williams to convince him to do so.[51]

Smuts's overriding preoccupation at this time was the extension of South Africa's boundaries and influence as far as possible. His vision of a greater South Africa saw the development of important economic and political bridges between the Union and all European-settled areas as far north as Kenya. The first step was to convince Southern Rhodesia to join the Union, provision for which had been made in the 1910 Union of South Africa Act. In 1915 a supplemental charter had extended the British South Africa Company's governing powers until October 1924 unless before that time an absolute majority of the Southern Rhodesian Legislative Council (LegCo)

[48] Grahame to Curzon, No. 270, 13 Apr. 1922, and No. 282, confidential, 19 Apr. 1922, FO 371/8238/265.

[49] TFAAC to Treasury, 10 Mar. 1922, T 160/347(3)/4273.

[50] Treasury to CO, urgent, 11 Mar. 1922, CO 551/155.

[51] Minute by Masterton-Smith, 3 Apr. 1922 and note thereon by Bottomley; Lord Buxton, Governor-General of South Africa, 1914–20, to Williams, very private, 3 Apr. 1922, copy in Buxton to Masterton-Smith, 3 Apr. 1922, CO 551/157.

could convince the British government that the territory was in a position to govern itself successfully. Chartered had wanted the two Rhodesias to amalgamate, but most of the members of the LegCo were even more opposed to this than they were to the alternative idea of becoming part of the Union. A movement for Responsible Government began to develop, receiving considerable impetus from the decision by the Privy Council's Judicial Committee in July 1918 that the unalienated lands in Chartered territory belonged not to the company but to the Crown. As long as Chartered administered the territories they were allowed to use revenues from these lands to reduce the administrative deficits they had accumulated over the years. Chartered applied to the government for an immediate refund of these deficits and told the Southern Rhodesians that, as the unalienated lands could no longer be regarded as a commercial asset, the company would invest no further capital in them. Lord Milner, who had become Colonial Secretary in 1919, in order to delay a final decision, appointed a commission under Lord Cave to determine the amount of money to which Chartered was entitled from the Crown. After the Cave Commission reported in January 1921 the Southern Rhodesia Legislative Council voted, by a large majority, a resolution favouring Responsible Government. Winston Churchill, who had replaced Milner in February 1921, was more sympathetic to the resolution and appointed another Commission, this one under Lord Buxton, to consider the possibilities of implementing it. The Buxton Commission recommended that if a referendum held in Southern Rhodesia showed that a majority of voters wanted responsible government, Southern Rhodesia should become a Crown Colony rather than a Protectorate.

Chartered and other mine owners wanted Southern Rhodesia to join the Union; they felt this was the only way financial stability could be maintained once Chartered withdrew its support. There were many traditional, legal, and economic justifications for such incorporation, they argued, but among Southern Rhodesian settlers, opposition to the idea continued to grow. People were afraid of centralized rule from bilingual Pretoria, and the threat to their continued existence as a British community which would be posed by the Afrikaaner population they would be joining. Many European workers feared competition from an influx of 'poor Dutchmen' and from a policy they expected to see introduced of increasing the use of cheaper African labour to replace Europeans, a fear which was hardly supported by increasing job reservation in the Union. In South Africa itself there was a more than considerable body of opinion opposed to bringing Southern Rhodesia into the Union. Many of Smuts's opponents who had acquired additional support when South West Africa, as a Mandated Territory, became an administrative part of South Africa, accused him of wanting to bring Rhodesia in merely to counterbalance their increasing strength. At about the same time as this question was coming to a head, a Portuguese delegation arrived in South Africa to begin negotiations for the revision of the 1909 Moçambique Convention regulating the joint use of the Port of Lourenço Marques (Delagoa Bay) and the supply of labour from Moçambique for the Transvaal mines. In the face of decreasing electoral support, Smuts thus had several very important questions to deal with at the same time.[52]

A delegation from Southern Rhodesia met with Smuts in Cape Town in May 1922 to discuss the terms he was prepared to offer if Rhodesia did decide to join the Union. Discussion included railway questions and the proposed guarantee. Sir Drummond Chaplin, Administrator of Rhodesia, was convinced that the completion of the Benguela Railway and the resulting deflection of traffic would have a disastrous effect on development in both Rhodesias.[53] Sir Charles Coghlan, leader of the Responsible Government movement and later Southern Rhodesia's first Prime Minister, associated the elected members of the LegCo with the view. They felt that any assistance the British government might be able to give could better be used to build a railway link between Sinoia, just north of Salisbury, to the railway line in Northern Rhodesia at Kafue, thus substantially shortening the route between Katanga and Beira. (A Sinoia–Kafue rail link was often mooted, but was never built because of the loss that would result for the line from Bulawayo to Kafue via Wankie.) As an alternative, government help might be applied to the construction of a line from Wankie through South West Africa to Walfisch Bay. Such a line was thought by some to be necessary to facilitate coal exports to South America.[54]

[52] cf. Sir W. Keith Hancock, *Smuts*, Vol. ii, *The Fields of Force* (Cambridge 1968), pp. 150–4; E. Walker, *A History of Southern Africa*, 3rd ed. (Longmans, London, 1957), pp. 592–9; Gann, . . . *Southern Rhodesia* . . ., pp. 231–46.

[53] Chaplin to Prince Arthur of Connaught, Governor of the Union and High Commissioner for South Africa, telegram, 20 Apr. 1922, in Prince Arthur to CO, telegram, 29 Apr. 1922, CO 417/678.

[54] Prince Arthur to CO, telegram, 2 May 1922, CO 417/678; Thomson to Williams, 20 Apr. 1922, TC 134.

Alec Thomson, Williams's nephew and representative in Katanga, went to South Africa to maintain contact between his uncle and Smuts. When he asked whether the conference on the Moçambique Convention had anything to do with the Prime Minister's attitude towards the Benguela Railway, Smuts embarked on a long discussion of sugar duties.[55] Smuts was almost undoubtedly using his opposition to the Benguela Railway at least in part as a lever against the Portuguese in his attempts to convince them to turn over the entire administration of the Lourenço Marques harbour to him. Williams became convinced that Smuts intended to take over the Portuguese colonies entirely, cloaking his activities in the 14 points set out by President Wilson at the Paris Peace Conference. Smuts would, Williams believed, work up interest and encourage Boer settlement on these colonies until enough Boers lived there to request protection from Portuguese misrule. Williams stoutly maintained that it was in Britain's interest to keep the Portuguese flag flying in Portugal's African possessions.[56]

Williams was certainly letting his imagination get the better of him. Smuts was hardly firmly enough in power to manipulate large numbers of Boers to the extent envisaged. The views Williams expressed were not, however, entirely without foundation. The British Consul-General at Lourenço Marques had earlier reported that one of Smuts's ideas regarding the revision of the Moçambique Convention seemed to be to find some way to get the Portuguese to cede to the Union the territory over which British reversionary rights had been secured accorded to the terms of the revised Anglo–German 'Secret' Treaty.[57] During the war there had been some interest expressed in the possibility of an exchange of territory in German East Africa for part of Moçambique in order to consolidate Union territory south of the Zambezi and Kunene Rivers. The Imperial War Cabinet refused in 1917 to put pressure on Portugal to cede Delagoa Bay as part of such a plan.[58] One of Chartered's directors said that he had learned that Smuts was prepared to absorb both Rhodesias and Nyasaland and 'freeze out Portugal'.[59] All these were ideas only and did not justify Williams's fears.

55 Thomson to Williams, personal, 11 May 1922, TC 134.

56 Williams to F. Swann (one of his ghost writers), 25 July 1922, Williams Papers.

57 H. Hall, Hall to Curzon, No. 4, confidential, FO 371/8368/202.

58 Hancock, i. 408, 498, 553.

59 Michell to Gell, private, 2 Sept. 1921, Gell Papers.

In 1923 Williams toured his various interests in Africa and had an opportunity to fill in for himself some of the details of what had been happening. He was told that Smuts had been very badly advised regarding his talks with Thomson as well as Southern Rhodesia. Coghlan said he entirely agreed that the Benguela line would get all Katanga's traffic, but for Rhodesia the Walfisch Bay route was the only economic possibility.[60] Williams found it rather amusing to hear Rhodesians say that the line to Beira was uneconomical for him while Smuts and Chartered were saying that the Beira line was the only reasonable one for both Katanga and Rhodesia traffic.[61] At lunch with Smuts Williams talked of some of his early experiences in Africa; conversation tended to remain general and fairly pleasant. At one point Smuts said to Williams, 'I suppose you will gradually annex these countries in the North?' Williams replied, 'Oh no, but that is just the great fear these peoples have that *you* mean to do.' Silence followed.[62] Another report indicates that between Williams and Smuts 'the hatchet had not yet received decent burial': replying to a question from Smuts Williams confirmed that he had already raised £9,000,000 for constructing the Benguela Railway, and 'would have got another £5,000,000 if the Prime Minister had not put a spoke in the wheel'.[63]

In September 1923 Prince Arthur visited Katanga. His secretary, Captain Clifford, discussing the Benguela Railway with Thomson expressed what Thomson assumed was the view of the Union Government: the Benguela Railway was inevitably going to be built and it was therefore preferable that it remain in British hands. Clifford strongly urged Thomson to cable Williams in England suggesting an interview with Smuts who was also in London at the time, and then a new approach to the Trade Facilities Act Advisory Committee. As in November Southern Rhodesians had rejected the idea of joining the Union by a majority of 2,785 out of 14,763 votes cast, Clifford thought Smuts's imperialistic views were now bounded by the interests of the Union alone.[64]

[60] Williams's Diary, 4 Apr. 1923.

[61] Ibid. 5 Apr. 1923.

[62] Ibid. 15 Apr. 1923.

[63] Prince Arthur to the Duke of Devonshire, Colonial Secretary, 1922–4, confidential, 24 May 1923, CO 532/239.

[64] Thomson to Williams, confidential, letter and telegram, 10 Sept. 1923, TC 134; Gann, . . . *Southern Rhodesia* . . ., p. 247.

Clifford was wrong, and it is impossible to understand why Williams should have regarded this as a direct communication from Smuts.[65] There is no reason why Smuts, who could easily have contacted Williams in London, should have used Clifford—the Imperial Secretary—to transmit his views. Williams did have some other reasons for believing that Smuts might have changed his mind, one being that he had been told in Cape Town that if he had seen Smuts three weeks after the 'veto' had first been imposed, Smuts would immediately have withdrawn his opposition. Chaplin had also assured Williams that Smuts would no longer bother about the Benguela Railway.[66] They too were wrong.

Before approaching the Advisory Committee again, Williams, with the help of Viscount Grey, met the Duke of Devonshire, who had replaced Churchill in October 1922, to try to ascertain the Colonial Office position. The only new point he raised was that following Smuts's veto of the guarantee, two Belgian banks had acquired a substantial interest in Tanks by taking up preference shares they had had to issue. This meant that orders for material and equipment would be placed in Belgium rather than Britain unless the government decided to assist finance.[67]

Smuts had not changed his mind at all. While he was generally in sympathy with any attempts to develop the African interior and agreed that the Katanga mines were of great value, he remained convinced that the Belgians were embarrassed by their obligation to build their section of the line, and intended to press on with the BCK. Once that was completed he felt the Benguela Railway would become virtually useless as it passed through an area which would supply very little supplementary traffic of its own. He also appreciated the political difficulties which could arise from British government support for investment of further British capital in a railway passing through two foreign colonies to the detriment of railway undertakings in British territory. He was still opposed to the guarantee, and Masterton-Smith assured him that the Colonial Office, acting on his and Rhodesia Railways' behalf, would continue to oppose it as

[65] Williams to Thomson, private and confidential, 1 Nov. 1923, TC 134.

[66] Williams to Wingate, 20 May 1923, SA 241/1; Williams to Thomson, confidential, 14 Nov. 1923, Williams Papers.

[67] Grey to Devonshire, 21 Sept. 1923 in Williams to E. H. March, Devonshire's Private Secretary, 22 Sept. 1923, CO 532/250; Memorandum by Hutchinson, 19 Oct. 1923, in CO 532/269. See below, pp. 109–18.

well. Clifford may have been won over to Williams's side; Smuts was not.[68]

It was only after Smuts's defeat in the 1924 elections in South Africa, and his replacement as Prime Minister by General Hertzog, that Williams was able to negotiate a British government guarantee for debentures successfully in order to complete construction. As soon as Smuts vetoed the guarantee and it seemed clear that he was not prepared to change his mind, Williams had put in train other schemes to raise the money to complete the railway, retaining 'control' over it. He was unable to regain the share of Union Minière he had already lost.

[68] 'Note for Record' by Masterton-Smith, 10 Oct. 1923, and minutes thereon, CO 532/270.

A complete understanding of Smuts's position would require more knowledge than is available about the sources of financial support Smuts had for his political activities, and his involvement with South African financial interests. Neither Sir Keith Hancock, Smuts's latest biographer, nor Sir Theodore Gregory, Sir Ernest Oppenheimer's, shed much light on this. (Hancock, *Smuts*, and Gregory, *Ernest Oppenheimer and the Economic Development of Southern Africa*) (Oxford University Press, Cape Town, 1962.) Smuts's son merely says that 'Many old friends, who all prefer to remain anonymous, . . . came to the Party's assistance . . . It was all done modestly and discreetly and the public will never know about the services of these benefactors' (J. C. Smuts, *Jan Christian Smuts* (Cassell, London, 1952), p. 328).

CHAPTER VII

End of the Line

UNTIL the British government changed their mind about supporting the Benguela Railway, Williams tried several ways of raising money, with only limited success. In the process his relations with his Belgian associates deteriorated still further, while continuing political instability in Portugal made it virtually impossible for him to make definite agreements with the government or to secure any help from them. By the time the line was finally completed, Belgian policy in the Congo, the general world economic situation, and particularly the market for copper, combined to keep the Benguela Railway Company in serious financial difficulties for many years.

THE BELGIAN SHARE IN TANKS

In 1922 when Tanks increased its authorized capital to £3,500,000, a series of £750,000 10 per cent preference shares was created. Sir James Dunn, founder of the London banking firm, Dunn, Fisher & Co., and Alfred Loewenstein, a Belgian financier, took an option of £500,000 of the issue. Loewenstein was an interesting person, having in the early stages of his career managed to raise the money to pay off his late father's substantial business debts, and gone on to become involved in a number of hydroelectric projects, the manufacture of artificial fibres, and various other enterprises. In the process he acquired a not entirely unblemished reputation, and even one of his admirers admitted that he was a rather flamboyant character who was not completely trusted, particularly in England.[1] Jadot's opinion of him was not very high, although probably not entirely unbiased as Loewenstein had a connection with the Banque de Bruxelles, something of a rival to the Société Générale. When relations between Loewenstein and Williams had come to an end, Jadot volunteered the information that Loewenstein had a very unsavoury reputation in

[1] Hubert Meredith, *The Drama of Money Making: Tragedy and Comedy of the London Stock Exchange* (Sampson Lowe & Marston, London, [1931]), pp. 297–306.

Brussels, having been involved in some business affairs that were not very highly thought of, and having several lawsuits on his hands.[2] It has never been satisfactorily decided whether the fall to his death from his private aeroplane over the English Channel in 1928 was in fact an accident.

In January 1923 Dunn and Loewenstein took up 100,000 preference shares, their option on the remainder being extended, and agreed to loan Tanks up to £100,000 if asked to do so.[3] Loewenstein then accepted a seat on the Tanks Board.[4] He could not take it officially until the number of directors was increased by a shareholders' meeting, but was informed of the Board's activities, and attended their meetings informally. It was while doing this that he became directly involved in Williams's attempt to secure support for a debenture issue from the London merchant bank, Schröder & Co.

This was initiated through family connections. In 1920 Williams's younger daughter had married Col. Francis Follett, brother-in-law of Frank Tiarks, a senior partner in Schröder's. In August 1922 Tiarks, through Follett, offered to help find the money to complete the line. The £5,000,000 Williams had said he wanted was 'nothing' provided the security for it was good.[5] When detailed negotiations began Baring Brothers were asked to participate but refused because of their fear that competition from the BCK would make the line unprofitable.[6] (The fact that one of Baring's directors had a brother on the Chartered Board may or may not have been a factor in arousing this fear.) Tiarks and his associate, Major Albert Pam, suggested that as Chartered had 'good friends' in the City it might be a good idea to work through them. Williams said he was quite happy to be friendly with Chartered 'as he [had] always been' and as evidenced by the recently concluded agreement for Rhodesia Railways to receive £2 per ton for up to 30,000 tons of copper carried over the completed Benguela line. On the other hand he maintained that if Chartered were permitted to have any voice in determining Benguela Railway policy, that railway would suffer while the lines in Rhodesia would benefit. Tiarks agreed, but wanted to give Chartered a last chance to co-operate and subsequently

[2] Williams's Diary, 9 July 1923.
[3] 'Memorandum of interview . . .', ibid., 3 Jan. 1923.
[4] Loewenstein to Williams, personal, 14 Jan. 1923, Williams Papers.
[5] Follett to Williams, 20 and 28 Aug., and telegram, 29 Aug. 1922, BR 21.
[6] Schröder Wagg Précis Sheet.

blame them if British interests lost control of the Benguela Railway.[7]

Family feelings did not prevent the attempted driving of a very hard bargain which Williams found entirely unacceptable. He insisted on being given some assurance that the 900,000 railway company shares Schröder wanted would not find their way into foreign hands. If Belgians or South Africans held them it would indicate, he claimed, unfaithfulness to the Portuguese.[8] Such 'unfaithfulness', had it facilitated the completion of the line, would undoubtedly have been more than welcome to the Portuguese, but at the same time would have weakened Tanks's own position, clearly Williams's prime concern. Williams was particularly disturbed by the condition Schröder had included in his proposal that construction on the Benguela line could be stopped by him at any time the Belgians stopped work on their section. This would mean that Tanks would be pawning all their assets for a period of time which could be determined by the Belgians who could delay construction, write off depreciation at an accelerated rate, and gradually lower Tanks's returns from Union Minière so that they would have to sell their Union Minière shares in order to pay railway debenture interest. Williams showed that he did in fact have some understanding of the grounds for conflict between himself and the Belgians by pointing out that Tanks was being asked to agree to Schröder's terms as if all necessary arrangements had been definitely concluded for the Benguela Railway to carry large amounts of copper traffic; there was no guarantee that the Belgians would in fact complete their section of the line by 1926 (then the time limit for completing the Benguela Railway), despite the terms of the Pooling Agreement, although he remained confident that the BCK could never successfully compete with the Benguela line because of grades and curves and its great length.

Williams now said that construction of the line itself, without the provision of rolling stock, necessary modifications of the existing lines, or pier work at Lobito, could be completed for £1,300,000 which could be raised by a further issue of Tanks preference shares.

[7] Memoranda, 'The Surrender', 30 Nov. 1922, and two undated memoranda in the Williams Papers.

[8] The following paragraphs are based on: undated memorandum, 'If we accept S & Co's terms', BR 21.

Alternatively, on the basis of two tenders from contractors other than Pauling, all the work could be done and equipment supplied for £2,350,000, very much less than the sum anticipated by Pauling's estimate, which could be covered by a debenture issue of only £3,000,000 by Schröder. He saw no justification whatever for Tanks to provide £1,650,000 profit for contractors and financiers if the work could be limited or done more cheaply; Tanks could and should 'paddle [their] own canoe'. Who provided the lower cost estimates is not specified, and it is impossible to take these suggestions seriously for several reasons. The Portuguese government was most unlikely to agree to any plan which did not provide complete facilities, and if these proposals were serious, why did Williams not carry them through?

Dunn delayed agreeing definitely to conditions Tanks set for once again extending his option on preference shares, claiming to be trying to get Loewenstein to relieve him of part of the financial burden of taking it up. Loewenstein for his part maintained that the drop in the price of Tanks shares on the open market made it more difficult for him to take up further shares than would have been the case earlier as there would be a delay in introducing the shares on to the Brussels Bourse.[9] He thought that Williams regretted not having accepted earlier proposals he and Dunn had put forward, and was sure a better bargain could be struck when negotiations were resumed directly with Williams, who was out of London at the time. In the meantime Loewenstein intended to pursue his investigations of the Benguela Railway's position by discussing it with Franck and Odon Jadot of the BCK. He intended not only to keep Dunn informed of what he learned but also to pass on to Schröder any interesting information he might receive.[10]

Franck apparently said that the Belgians did not want to build their end of the line and, if forced to do so, would require financial assistance.[11] By relaying this 'interesting information' Loewenstein effectively brought negotiations with Schröder to an end. Williams's plans for financing his railway once again foundered on the rocks of strong doubt that it would ever become a paying proposition. With the end of negotiations came an end to amicable relations between

[9] Loewenstein to Scotland, 28 May 1923, Williams Papers.

[10] Loewenstein to Dunn, 28 May 1923, enclosed, apparently by mistake, in Loewenstein's letter to Scotland of same date, Williams Papers.

[11] Hutchinson to Sir Francis Wingate, 12 June 1923, SA 241/1.

Williams and Loewenstein. Williams refused to allow Loewenstein to take his seat on the board as his action in talking to Schröder while being privy informally to the proceedings of Tanks Board was a breach of confidence;[12] a highly irregular action which indicated, as far as Williams was concerned, that Loewenstein and Dunn were trying to eliminate Schröder from railway finance, get into it themselves by securing the contract to complete construction, and ultimately to gain control of Tanks and thereby a large interest in Union Minière.[13] He may well have been correct.

Loewenstein, although he and Dunn withdrew their offer to take up the remaining shares on which they had an option, tried to put a colleague, one Heinemann, on the Tanks Board in his place, an attempt which some French and Belgian newspapers reported as successful. Jadot said that Heinemann was a very clever, capable, and strong man, of quite different character from Loewenstein, but he was a personal enemy of Francqui and was connected with a group that as early as 1911 had tried to secure an interest in Tanks in order to get into Union Minière. Jadot also offered to give Williams any financial assistance which might be needed.[14]

An opportunity to make good this offer was not long in coming. Loewenstein offered the £227,000 preference shares he already held to Felicien Cattier, a Director of the Banque d'Outremer which was an affiliate of the Société Générale, on condition that Cattier support his claim to a seat on Tanks's Board. Cattier refused to accept this condition, feeling that Loewenstein wanted to acquire a connection with a reputable bank only in order to improve his standing in the financial world. Cattier was willing to support Williams generally, though not unquestioningly, and insisted that any agreement with Loewenstein would have to include a provision that Williams himself would not be adversely affected by it.[15] Although Williams expressed his appreciation of Cattier's position, he remained aloof from these negotiations, being convinced that Loewenstein would not press his point. If the Banque d'Outremer refused to buy his shares, becoming the second bank to do so, Loewenstein's reputation would remain what it was and he would have great difficulty finding anyone else to take them. The negotiations then in progress for a subsidy from the

[12] TCL to Loewenstein, 28 Sept. 1923, copy in SBK.
[13] Memoranda, Diary, 3 and 25 July 1923.
[14] Diary, 12 July 1923.
[15] Memoranda, 19 and 20 Sept. 1923, TC 151.

Angola government appeared most promising, and perhaps justified Williams's indifference.[16]

Loewenstein and Cattier reached agreement in October 1923 for the Banque d'Outremer to take over his shares.[17] Logically, part of the deal was that Cattier or one of his nominees was to have a seat on the Tanks Board. Tanks also decided to offer a seat to Jadot personally;[18] Jadot refused, asking that it be given instead to his assistant, Firmin Van Bree. To this request Tanks took strong exception.[19] The argument on this point went on for several months and became at times quite acrimonious. Williams claimed to have nothing against Van Bree personally; this was not strictly true as he told Lane that Van Bree was unacceptable because he had caused considerable trouble by introducing South African capital into the Angola Diamond Company. Cattier pointed out that the refusal to offer the seat to anyone but Jadot created a very bad impression in Brussels, to which Williams replied that if such was the feeling because Tanks had wanted to honour Jadot and his long association with them, the Belgian attitude could only be seen as antagonistic; if war was wanted, war there would be.[20] Williams was hardly in a position to make such threats, and not surprisingly the mailed Belgian fist came out of the velvet glove: if the second seat was refused the Société Générale could not continue financial support for Tanks. On a friendlier note, however, Jadot agreed not to press the matter, but remained convinced that Williams would ultimately agree that a Société Générale Director would strengthen the Tanks Board.[21]

Another source of friction was the generally unco-operative Belgian attitude towards the Benguela Railway, illustrated particularly by the press campaign being carried out against it and Williams personally. This and the lack of any public contradiction from Jadot was most damaging, and Williams pointed out that if large numbers of Tanks shares came on to the open market as a result of this campaign, Union Minière itself might suffer. Jadot defended himself by saying that he could not control all aspects of the question. In so

[16] Memoranda, 19 and 20 Sept. 1923, TC 151.
[17] Hutchinson to Williams, 16 Oct. 1923, TC 151.
[18] Diary, 25 Sept. 1923.
[19] Williams to Hutchinson, telegram, 17 Oct. 1923, and Cattier to Hutchinson, 23 Oct. 1923, TC 151.
[20] Diary, 19 Nov. 1923.
[21] 'Memorandum of Interview . . .', 26 Nov. 1923, TC 151.

far as Belgian obstruction was delaying the railway's completion, it was Franck Williams had to convince, and Franck was at the time rather preoccupied with impending elections. The railway, Katanga, and the BCK were part of some very delicate political issues with which Franck had to cope, and there was increasingly strong feeling in Belgium that the completion of the BCK was all that mattered.[22] None the less, four days after this conversation took place, an article appeared in the Brussels press praising Williams as a friend of the Congo, and putting forward the view that it was important for Katanga to have available to it as many transport routes as possible. Williams had, obviously with Jadot's help, succeeded in 'allaying the suspicions of his plans which were only recently very much in evidence in the Belgian press'.[23]

Williams was telling anyone who asked him at this time that finance for the railway had been arranged, that agreements had been reached for options to be taken up on the remaining preference shares, and that Tanks was taking 5 per cent railway debentures at 50 per cent. This was true as far as it went, but neither the debentures nor Tanks preference shares could produce all the money needed. Williams now sought the help of the oft-threatened American financiers.

By 1923 the Copper Exporters Association's effective life had come to an end, primarily because the Guggenheim-controlled members of it withdrew, but also because of excessive interest in its activities being shown by the United States government.[24] This did not mean that all American copper interests were no longer concerned about controlling the world price of copper and competition from Union Minière. In January 1923 a representative of Morgan's parent company in London, Morgan Grenfell, suggested that the Anglo–American Corporation might be willing to finance completion of the Benguela Railway. Williams did not want South African capital, but in March 1924 decided to go to America ostensibly to discuss possible Union Minière co-operation on price maintenance with some of Morgan's associates. He was not acting officially as Union Minière representative and was not authorized to conclude any agreements. Jadot knew Williams was making the trip but was 'not too keen' on it.[25]

[22] 'Memorandum of conversation . . .', Diary, 25 Feb. 1924.

[23] C. Wingfield, chargé d'affaires in Brussels, to MacDonald, No. 102, FO 371/10528/993.

[24] Skelton, pp. 415–27, and Herfindahl, p. 95.

[25] Diary, 12 Mar. 1924, and 'Memorandum', 21 Mar. 1924.

In the course of what were essentially meaningless talks with John Ryan of the Anaconda Company and Thomas Cochrane of Morgans, Williams said that the only way Ryan could secure any influence on Union Minière policy was to take an interest in Tanks; a seat on the Board could easily be found for him.[26] When Williams boasted that he 'generally got his own way with the Belgians', Ryan reminded him that Morgans had their own way of exerting influence.[27] Another advantage of an interest in Tanks, according to Williams, would be that the Benguela Railway ran through fertile, untouched country that could be developed as a food supply for Katanga.[28] Rather strangely he said he could not discuss railway finance because of negotiations with Schröder.[29] As Schröder was in New York at the time, it is possible that Williams made a new approach to him, but there is no evidence of this; Williams's statement is all the more incomprehensible as Ryan never showed more than a polite interest in the Benguela Railway and certainly no desire to finance it even when Williams openly admitted that this was what he wanted.[30] Nothing came of these talks, although they may have had some effect on consequent talks between Ryan, Jadot, and Francqui, in Union Minière's decision to participate in Copper Exporters Incorporated, founded in 1926 to take over the regulating function of the defunct Association, and in there being some American participation in Union Minière's capital increase in order to finance further expansion.

In order to put the full-scale leaching plant into operation, build a hydroelectric installation on the Lufira River, and construct more smelters, the company decided to issue further shares. Williams agreed, after some delay, to participate in financing the expansion, but was being forced to sell Tanks's rights to the privileged shares that were part of the scheme in order to raise needed cash.[31] At about the same time Union Minière decided to go ahead with the erection of a refinery at Hoboken, Belgium, using the proceeds of the sale of radium extracted from pitchblende deposits first discovered at Chinkolobwe in 1913 and successfully worked from January 1923.[32]

[26] Diary, 11 Mar. 1924.
[27] Ibid., 23 Mar. 1924.
[28] Ibid., 14 Mar. 1924.
[29] 'Memorandum . . .', Diary, 21 Mar. 1924.
[30] Diary, 25 Mar. 1924.
[31] Ibid., 20, 21, and 24 Nov. 1924.
[32] Ibid., 27 May 1924, and Union Minière Director's Report to the Annual General Meeting of 9 July 1923.

(These same deposits were the source of the uranium used in the first atomic pile built in Chicago, and in the atomic bombs dropped on Hiroshima and Nagasaki.)

The discussions with the Portuguese that Williams had thought quite promising while Cattier was negotiating to take over Loewenstein's preference shares were in fact with Norton de Mattos, who had been hailed as Angola's saviour when he took up his post as High Commissioner there in 1920. Norton de Mattos tried to secure portions of various loans which the Portuguese were trying to raise for the company.[33] Even if any of these attempts had been successful, Norton de Mattos was not in a position to give Williams any help at all, as he was no longer High Commissioner.[34] At best he was in a weak political position and could probably not have secured any support in Lisbon for anything he wanted to do. Feeling there was running very strongly against him and he was generally described as having been very corrupt in his colonial post and accused of misappropriating funds. Captain Francisco Pinto da Cunha Leal, who served briefly as Prime Minister in 1921 and had been a leader of the revolution of 1910, called him a 'tyrant, murderer and singularly corrupt administrator'; further charges against him were that he had unjustifiably stopped the publication of newspapers in Angola, forced Protestant missionaries out of the country, and shot his own horse 'as an example to other dumb animals'.[35] When Angola finances were discussed in the Cortes he was accused of having left the province's administration in a state of chaos.[36]

Much of the criticism of Norton de Mattos was politically inspired. He aroused strong opposition by his relatively enlightened labour policy and his attempts to eliminate forced labour, particularly for private purposes. He achieved much in the building of roads, schools, and hospitals, but only at great cost. Far from being a saviour, his 'one-man colonial rule . . . brought the province to the brink of bankruptcy'; while he was in office, Angola's debt increased tenfold. His attitude towards labour did not prevent his establishing a system of repression which crushed any attempt on the part of educated

[33] 'Memorandum of interview . . .' 21 May 1924 and 'Note of interview . . .' 26 May 1924, BR 21.

[34] Williams to Lane, telegram, 3 June 1924, BR 21.

[35] J. Duffy, *Portuguese Africa* (Harvard University Press, Cambridge, Mass., 1959), p. 252.

[36] Note by Lane, undated, but about 8 July 1924, BR 21.

Angolan Africans to have a voice in their own government.[37] He was forced to resign in 1924.

No replacement for Norton de Mattos as Angola's High Commissioner was appointed for some time. The Zambesia Exploring Company was able to provide the money to send out another 100 kilometres of permanent way material to keep work going,[38] and some hope for Portuguese assistance for the railway appeared to be in the offing when, in August 1924, Rego Chavez was appointed High Commissioner.[39] He seemed well disposed, at least initially, to the railway company and wanted to settle a number of questions concerning it as soon as possible.[40] A fundamental condition he imposed on his acceptance of the post was that sufficient funds be put at his disposal to enable him to fulfil his duties, particularly with regard to the problem of transferring funds from Angola to Europe. This was particularly important for the railway as Tanks had been forced to find cash in Europe to service railway debentures while large amounts of money, accumulated working surpluses which at the pre-war rate of exchange would have covered most if not all of these costs, remained inaccessible and useless in Angola. Chavez hoped that a £500,000 loan he had arranged with the Angola Diamond Company would provide sufficient hard currency to solve the problem.[41] Nothing definite could be done about this in time to be of any use to Williams. Events in South Africa opened the way for the last and ultimately successful attempt by Williams to secure the money needed to complete the line.

SMUTS'S LOSS IS WILLIAMS'S GAIN

In June 1924 the combined forces of the Nationalist and Labour Parties subjected Smuts and his South African Party to a resounding defeat, gaining a majority of 27 in the Assembly. Smuts, the 'sticking point' in Williams's attempts to secure British government support, was himself defeated in his constituency, opening the way for

[37] Duffy, p. 250; H. V. Livermore, *A New History of Portugal* (Cambridge University Press, Cambridge, 1966), p. 239; Basil Davidson, *In the Eye of the Storm: Angola's People* (Longmans, London, 1972), pp. 123–9; Douglas L. Wheeler and René Pelissier, *Angola* (Pall Mall Press, London, 1971), pp. 114–15.

[38] Appendix attached to an undated memorandum by Williams, BR 21.

[39] Lane to Williams, 3 Sept. 1924, BR 21.

[40] Lane to Hutchinson, 10 Oct. 1924, BR 21.

[41] Lane to Williams, confidential, 4 and 25 Nov. 1924; Williams to Lane, confidential, 18 Nov. 1924, BR 21.

Williams to ask Earl Athlone, who had succeeded Prince Arthur as Governor of South Africa in January 1924, whether or not there was now any objection to the British government's giving the Benguela Railway a guarantee of debenture interest.[42] There were still some objections, but Hertzog, the new Prime Minister, was less imperialistic and less imperious than his predecessor, and did not press them. Although not particularly in favour of the Benguela Railway, he did not want to do anything to annoy the Portuguese, as he was much more concerned about concluding a new Moçambique Convention than with Britain's wider imperial aims.[43] Williams then asked the Trade Facilities Act Advisory Committee if they were prepared to resume negotiations. Since discussions had ceased, a further 200 kilometres of line had been laid and the financial position of all the companies involved considerably improved. Union Minière was paying regular dividends and Tanks's interest in that firm had more than trebled in market value.[44] If the British government continued its refusal to give the guarantee Williams went on to say that he would be forced to buy materials from foreign suppliers who quoted prices considerably lower than those of British manufacturers.[45]

Before replying the Committee asked the Colonial Office if there had been any change in their view of the matter. Rhodesia Railways and the government of Southern Rhodesia continued to oppose the guarantee on the grounds that the British government should not assist a foreign railway which would damage Rhodesia's own lines. Union Minière's copper traffic provided more than one-third of Rhodesia Railways' gross receipts, half their net revenue. The loss of this traffic would mean, they claimed that Southern Rhodesia's own import and export traffic would have to bear higher transport costs which would retard the country's development. Hutchinson pointed out that minimum traffic guarantees had already been given to Rhodesia Railways in order to insure adequate supplies of coal and coke for the mines, and the Committee remained convinced that the Benguela line would be completed in any case, but that without a

42 Ulick Alexander to Williams, 17 Dec. 1924, Williams Papers.

43 'Note by the Secretary of State for the colonies', 6 Mar. 1925, Cabinet Paper No. 140 (25), CAB 24/172.

44 Williams to TFAAC, 16 Jan. 1925, TFN.

45 TFAAC to CO, undated copy in CO to FO No. 6336, 16 Feb. 1925, FO 371/11042/1273.

government guarantee, orders for supplies would be placed outside Britain.[46]

The point on which the Committee still had to be convinced was whether or not the Belgians could be counted on to honour their commitment to link the completed Benguela Railway to the mines. Henry Carton, who had replaced Franck as Colonial Minister in 1924, said he was convinced that the line should be built, but did not consider it particularly urgent. To Williams's argument that flooding on the Beira line delayed traffic and caused substantial loss, Carton replied that Williams exaggerated the seriousness of the problem and that in so far as the Beira line was inadequate, the rail/lake link to Dar-es-Salaam provided an adequate means (at least for the time being) of relieving the difficulty. The BCK was not mentioned, though it clearly was very much in Carton's mind. Before committing themselves, however, the Belgians wanted to be convinced that the Benguela Railway would in fact reach the frontier. Williams was caught in a vicious circle: the Belgians would agree to fulfil their obligation when Williams could tell them when the Benguela line would reach the Katanga border. In order to secure the money to build the line to that point Williams had to have assurances that the Belgians would build their section. As a way out of the dilemma Williams accepted Cattier's suggestion that an agreement supplementing the Pooling Agreement be drawn up laying down the terms under which the Belgians would build the link.[47] This was not, in the end, done. Although the Committee would not commit themselves until this point was settled, it was obvious that opposition to the guarantee was weakening.

The Board of Trade said that as the line would be built whether or not the government assisted, Rhodesia Railways would not suffer from the government guarantee any more than they would in any case. The government could no longer be accused of sacrificing Rhodesian interests to the British iron and steel industry which needed as much help as it could get. The Colonial Office, though deprived of the Union government veto fought a rearguard action, arguing that the railways' objections were well founded and in themselves sufficient reason for refusing the guarantee. They also raised what they con-

[46] TFAAC Memorandum, 3 Feb. 1925, and Newton to TFAAC, 6 Feb. 1925, copies in CO to FO No. 6336, 16 Feb. 1925, FO 371/11042/1273.

[47] 'Memorandum of interview . . .', Diary, 26 Feb. 1925.

sidered the constitutional problem of doing something opposed by a self-governing colony that gave substantial preference to British imports. The guarantee would not violate the clause in the Letters Patent giving Southern Rhodesia responsible government that had reserved legislation on railway matters to the imperial parliament, but would certainly be wholly inconsistent with its protective intent.[48] Although Williams did not provide details of the alternative finance available to him that would have involved placing orders outside of Britain, there are numerous references in various memoranda to the willingness displayed by Van Bree and others to arrange for the Société Générale to do this. Belgian interests were being fully protected by the construction of the BCK. Further assistance to Williams would give them greater influence over his companies and activities than they already had, while lessening still more Williams's ability to exert any pressure on them. Not surprisingly Williams preferred to give as little to the Belgians as possible, and because of the apparent certainty that the line would be completed whatever the British government did, the Colonial Office argument carried very little weight.

Before making a final decision the Cabinet wanted the views of the two Rhodesias.[49] As expected the North was very much in favour of the guarantee: they hoped that eventually a link between the Benguela Railway and the Copperbelt, the serious exploitation of which was still in its earliest stages, would be built. The expansion of activity in Katanga that the Benguela Railway would promote was still expected to redound to the advantage of Northern Rhodesian farmers. (This was in fact not as certain as it had been earlier as Southern Rhodesian farmers were beginning to invade the Katanga market by this time.) Southern Rhodesia said they did not want to obstruct anything that would help ease the unemployment problem in Britain, but went down fighting with the final shot that the Benguela Railway would 'introduce a foreign competitive factor into carrying trade hitherto enjoyed by the railways' of Rhodesia.[50] The Cabinet then agreed

[48] Cunliffe-Lister, President of the Board of Trade, to L. S. Amery, Colonial Secretary, 29 Jan. 1925, and to Walter Guinness, Financial Secretary to the Treasury, 20 Feb. 1925, copies in T 160/547(3)/4273, and 'Note by the Secretary of State for the Colonies', 6 Mar. 1925, Cabinet Paper No. 140(25) CAB 24/172.

[49] Minutes of Cabinet Meeting, 11 Mar. 1925, CAB 23/49/15.

[50] 'Replies from the Governments of Northern and Southern Rhodesia', 1 Apr. 1925, Cabinet Paper No. 192(25), CAB 24/171.

that in view of the certainty that in any event the Benguela Railway would be built; on the understanding that if a guarantee was granted orders would be given in [Britain] for railway material, that is to say, in the industry where unemployment [was] at its worst; and having regard to the tenor of the telegrams from the Governments of Northern and Southern Rhodesia; the Trade Facilities Committee should be free to consider an application for a guarantee on its merits.[51]

With political obstacles out of the way Williams and the Committee were able to agree on essential details of the guarantee in May 1925.[52] Although the African Railway Finance Company, the company formed to issue the debentures, was not established for more than a year, government anxiety to secure the guarantee's practical effects for British industry was such that Williams was asked to start placing orders almost immediately. He responded with alacrity, all orders being placed before the end of July.[53] To think that this cleared the way for the completion of construction would be, however, to forget the chaotic political situation existing in Portugal at the time.

The Portuguese government refused to extend beyond 1926 the time limit for finishing the line. This had been arranged with Norton de Mattos and confirmed unofficially by Chavez. After a private discussion with Estevan Pimental, one of the railway's government Directors, Chavez refused to make this confirmation official. Williams said that if the Portuguese continued he would simply tell the Committee the money would not after all be needed and construction would stop at the Cuanza River; this 'would undoubtedly shake everybody's faith in Portugal and damage her credit irretrievably'.[54] Whether or not anyone had any faith in Portugal left to be shaken is a moot point.

In an attempt to settle the matter Williams went to Lisbon where he refused to agree to suggestions that he should negotiate the time limit and possibly resolve some other outstanding differences between the railway company and the government. In his view the extension had been granted but then withdrawn; to negotiate would constitute an admission that he had been dealing with the Trade Facilities Act Advisory Committee under false pretences, which he had not done. Sir Lancelot Carnegie agreed with Williams's position and asked for

[51] Cabinet Minutes, 8 Apr. 1925, CAB 23/49/20(25).
[52] 'Memorandum . . .', Diary, 20 May 1925.
[53] Sainsbury (TFAAC Secretary), to Williams, private and confidential, 4 June 1925, and Williams to TFAAC, 22 July 1925, TFN.
[54] Memoranda, Diary, 10 and 22 Sept. 1925.

some 'considerata' that he could forward to the Foreign Minister, Dr. Vasco Borges.[55] It is interesting that Carnegie undertook to assist Williams without informing the Foreign Office or awaiting instructions from them. Carnegie, Williams, and Lane met with Borges who said that many people were convinced the Belgians would not build their portion of the line. Williams could only reiterate assurances that press reports claiming this were not true; if he had not been convinced the Belgians would honour their obligations he would not have asked his shareholders to continue putting up money for construction. He said he had come to Lisbon 'expecting an honour for all he had done for Angola and not to be asked to give something to Portugal!!!'[56]

Carnegie thought there might be some reason to believe that German interests might be behind opposition to the railway, because of the report he had received from the British Consul in Angola to the effect that the Banco de Angola e Metrópole which had recently begun operations there, was backed by German, not Dutch interests as had been claimed. Count Lichtervelde, Belgium's Ambassador in Lisbon, said that according to his information it was Russian money that was behind the bank.[57] Confusion as to the source of this bank's capital is quite understandable. It was formed as part of one of the more highly publicized Portuguese scandals of the 1920s—the 'Portuguese Bank Note Case'. Official Bank of Portugal stationery was used to instruct Waterlow & Sons in London to print 580,000 Portuguese 500-escudo notes, which were smuggled into the country in the Venezuelan diplomatic pouch. The bank was to dispose of the proceeds of the fraud, but had a very short life, being liquidated when the discovery of duplicate numbers on notes in circulation confirmed suspicions earlier aroused by the boldness of its activities.[58]

Apart from any specific difficulties created by this bank or fear of Belgian intentions, there was a general anti-Williams feeling prevalant in Lisbon, through no fault of Williams's own. For several months in 1924 an American sociologist, E. A. Ross, had visited Angola and Moçambique to investigate the welfare of Africans there. After a

[55] 'Diary of events and interviews in Lisbon from 5 October to 28 November 1925 in connection with extension of time for construction of Benguela Railway to 1928' (henceforth cited as Diary of Events), entries for 5 and 6 Oct. 1925, Williams Papers.

[56] Diary of Events, 9 Oct. 1925.

[57] Ibid., 13, 25, and 27 Oct. 1925.

[58] Livermore, p. 329.

visit totalling less than two months, during which time most of his inquiry was carried out in the vicinity of Protestant mission stations, he presented a report to the Temporary Slavery Commission of the League of Nations, although the Commission had not sent him to investigate in the first place. The report, highly critical of the Portuguese administration, was carefully rebutted by the Portuguese delegation to the commission, and no action was proposed. One part of the report, however, continued to affect feelings towards Williams, the 'Statement of Mr. F of a large contracting firm engaged on harbour works at Lobito Bay':

> This firm has at work 1,000 natives who have been obtained from the Government. It gives them a blanket and a jersey and they serve for six months. At the end of the term the firm gives them one-fifth of their wages and pays the remaining four-fifths to the Government which remits it to the *chefes de posto* where the men were recruited. We know that the money is actually transmitted to the *chefes* for we see their signatures, but we don't know what becomes of it then.[59]

Pauling was the contractor in question and there was considerable anger aroused by the fact that an employee connected with Williams, although indirectly, had given damaging and untrue evidence to someone like Ross who, it was claimed, had not even enjoyed the support of the American Consul in Angola and had been repudiated by the 'American Head Missionary' there. Mr. F turned out to be a Mr. Fletcher who had already been threatened with legal action if he continued to represent himself as being connected with Williams or the Benguela Railway, not the Mr. Freeton employed by Pauling. The report was in any case inaccurate as the men were paid monthly, not at the end of their contract.[60]

Many months after the furore had died down in Portugal the British Anti-Slavery and Aborigines Protection Society drew the Foreign Office's attention to the report and claimed that the Benguela Railway Company, Pauling, and Norton-Griffiths, who was carrying out harbour works at Loanda, were guilty of labour abuses. Pauling and the Railway were specifically castigated for making 'financial arrangements with Portuguese officials' in order to fill their labour

[59] E. A. Ross, *Report on Employment of Native Labour in Portuguese Africa* ([The Abbott Press], New York, 1925), p. 39; Duffy, pp. 166–8.

[60] Pauling & Co. to Scotland, 22 Sept. 1925, enclosing copies of sections of Portuguese Colonial Ministry Bulletins Nos. 2 and 6 of 1925, copy in Williams to FO, 16 Sept. 1926, 371/11136/1588.

requirements.[61] These charges disturbed Williams very much as he took a not unjustified pride in the excellent reputation he enjoyed with regard to African labour matters, and blamed long-standing enemies—his ubiquitous scapegoats—for the complaint. He also said that grave charges were about to be levelled against the company for using forced labour.[62] Everyone, including Williams, was over-reacting to what was little more than a storm in a teacup. The report from the Vice-Consul at Lobito satisfied the Foreign Office that the Anti-Slavery Society's charges were not justified.[63] The Foreign Office was also satisfied that the committee Williams sent out to investigate the situation was not entirely a 'white-wash job'. Their report could not have been as good as it was if reasonably satisfactory working conditions did not prevail. It was only at this point that anyone seemed to appreciate the fact that the only specific allegation against the three companies contained in the Ross Report was with regard to the remission of the four-fifths payment to workers. This really concerned only the Angolan government.[64]

By mid-October 1925 the Portuguese Cabinet agreed to the time extension on conditions that were generally unimportant and could easily be satisfied. One exception was their insistence that when the line reached Moxico the government would be able to determine the direction it was to be continued, presumably with the idea of deflecting it to the north of Angola or to the Northern Rhodesian Copperbelt. This was unacceptable as it would upset arrangements for the British government guarantee as well as the Pooling Agreement. Williams even refused to accept the minor conditions, continuing to maintain, at least openly, that the extension had already been granted. The Foreign Minister, who supported Williams in the Cabinet, suggested some political blackmail. If Williams looked into the Rumagoa Concession—just the mention of it would perhaps suffice—he might be able to silence opposition to the railway. This Concession had been granted in connection with a hydraulic power scheme for one of the rivers marking the Luso-Spanish frontier. The

[61] Anti-Slavery and Aborigines Protection Society to FO, 10 Aug. 1926, and Williams to Lane, strictly private and confidential, 13 Sept. 1926, in Williams to FO, 29 Sept. 1926, FO 371/11136/1588.

[62] FO memorandum, 27 Sept. 1926, attached to Williams to FO, 16 Sept. 1926, and Williams to Machado, 13 Sept. 1926, copy in Williams to FO, 29 Sept. 1926, FO 371/11136/1588.

[63] Minutes on CO to FO No. X.F. 7362, 12 Nov. 1926, and Hutchinson to FO, 3 Nov. 1926, FO 371/11136/1588.

[64] Minutes on Williams to FO, 17 Dec. 1926, FO 371/11136/1588.

two countries were at this time unable to reach a definite agreement between themselves, but were prepared to co-operate with a foreign group.[65] Proposals for straightforward bribery followed. Lane was told that some of the ministers opposed to the company could be 'squared' and later received an anonymous telephone call telling him that if the railway company did not pay a 'certain person' £5,000 there would be great resistance to the extension. The caller also said that the Foreign Minister was 'hard up and wanted some cash'.[66] One can only guess at the intricate political machinations behind these suggestions, and there is no evidence that Williams acted on them. It was only on 20 November after considerable struggle in a much-divided Cabinet that the government reached an acceptable decision.

The conditions of confirming the time extension were that the line reach Moxico within two years of the completion of the bridge over the Cuanza River, and the Congo frontier by the end of 1928. No further time extension would be granted. The future Portuguese government portion of any new railway company share issue would be 15 per cent, given them fully paid; when all shareholders had received a 5 per cent dividend, the government was to receive an additional 5 per cent; when dividends to all shareholders reached 10 per cent, the government was to get a further 7½ per cent. The railway company had to assume responsibility for repair and maintenance of the access roads to railway stations within the original land concession, and international traffic was not to be allowed to prejudice the transport of Portuguese goods. The final condition was that the government had the right to 'intervene opportunately' in the line plans beyond Moxico, taking the connection with the Congo into account.[67] The Colonial Minister advised the railway company Board to accept the terms, leaving formal approval by a meeting of shareholders until a later time 'as governments did not last long in Portugal' and any delay might mean negotiations would have to begin yet again. On 25 November 1926 the Lisbon Board signed the definite contract for the completion of the line.[68]

The only apparent problem remaining was to convince the Belgians to begin work on their portion of the line. This they were in no hurry to do, and despite constant pressure from Williams, the

[65] Diary of Events, 15 and 18 Oct. 1925.

[66] Ibid., 9 and 14 Nov. 1925.

[67] D. Pereira to the Government Commissioner of the BRC, 20 Nov. 1925, copy in Diary of Events, and Diary of Events, 22 Nov. 1925.

[68] Diary of Events, 22 and 26 Nov. 1925.

BCK did nothing to start construction. They were primarily concerned with the completion of their own line. The CFK had reached Bukama in 1918, and four years later surveys were begun in preparation for the linking of Bukama to Port Francqui (Ilebo) on the Kasai River. Actual construction began in 1924 and proceeded rapidly with relatively little difficulty, although there was considerable pressure on the country's labour resources. It was only in March 1927 that the BCK informed Williams that they were sending men out to Katanga to begin preliminary work on the link between the CFK and the Benguela Railway.[69]

THE BENGUELA RAILWAY IS COMPLETED

Construction of the last stages of the Benguela Railway also proceeded without hindrance. What had for more than forty years been accepted as the frontier between the Congo and Angola was reached on 27 November 1927, well before the time imposed by the Portuguese had elapsed.[70] Four months earlier, however, the frontier had been moved. The Belgians wanted to undertake improvement and expansion of the facilities at Matadi in order to cope with the increased traffic expected when the BCK was completed, and as the country's general economic development continued. The Congo's boundary with Angola near Matadi left very little room for growth; unless the boundary were changed, extensive costly tunnelling work would have been necessary. What the Belgians referred to as a 'slight rectification of the . . . frontier' by providing more space in less hilly terrain, would make the work much easier and cheaper. In May 1926 a suggestion that such a 'rectification' might be made was included in a semi-official memorandum sent to the Portuguese Foreign Minister putting forward ideas on a variety of possible areas of Luso–Belgian co-operation with regard to their African possessions.[71]

The only way it was politically possible for Portugal to agree to give up any of her territory at all, no matter how unproductive it was, was to receive in return territory 'several thousand times larger' than the three square kilometres of land Belgium wanted.[72] Preliminary discussions began in Lisbon in December, followed by more detailed

[69] Williams's Diary, 29 Mar. 1927.

[70] Williams to Machado, 6 Dec. 1927, Williams Papers.

[71] Carnegie to Chamberlain, No. 132 confidential, 22 May 1926, FO 371/11933/2126.

[72] Ibid., No. 260, 14 Oct. 1926, FO 371/11933/2126.

talks in Loanda. A series of four conventions emerged from these negotiations, two of which involved the Benguela Railway. The Convention of 21 July 1927 bound the Portuguese to equip the harbour at Lobito Bay adequately to cope with Katanga traffic, while the Belgians were bound to complete the rail link from the frontier to existing railways in Katanga within five years of the Benguela Railway's completion. Another convention signed the next day provided for Belgium to take over a small amount of land near Matadi in exchange for about 1,375 square miles west of Dilolo.[73] As a result the Benguela Railway had to be extended by about 100 kilometres. The new frontier was reached on 28 August 1928.[74] The line was officially opened with great ceremony the following June. At an average rate of 52 kilometres per year, it had been pushed across Angola and only awaited the building of a further 522 kilometres of line to Tenke to tie it to its major source of traffic, fulfilling its original purpose.

The BCK was completed to Ilebo in 1928 also, and was formally opened in the presence of the King and Queen of the Belgians. Earlier plans had envisaged the running of the rail parallel to the Kasai and Congo Rivers, joining the Bas–Congo Railway at Leopoldville. The completion of the line as far as Port Francqui made available a rail/river link that could serve as a viable alternative to both the Rhodesian and Benguela routes. It was not anticipated, however, that there would be too much difficulty for any of the lines as there was still great confidence that sufficient traffic would be generated for all railways serving Katanga to have a large amount to carry. The Benguela Railway was also protected by the provisions of the Pooling Agreement, but these proved useless. The first opportunity to revise the agreement had been ignored by all concerned in 1920. By the end of 1928, however, circumstances had changed so that the Belgians did not think there was as much advantage for them as they had originally thought. The main reason for this was that the CFK, which had been seen primarily as a temporary measure in 1908, necessary only until the Benguela and BCK routes were opened, had in fact proved highly profitable and essential for carrying coal and coke from Wankie. It had become an attractive investment, and the Belgian government was supporting an issue of privileged shares to finance

[73] Belgium, Chambre des Représentants, Documents Parlementaires, No. 12, 1927.

[74] TCL Annual Reports for years ending 31 Dec. 1927 and 1928.

improvements. Holders of these shares could not have been expected to be pleased to see working profits from their lines diverted either to the BCK itself or, even less so, to the foreign Benguela Railway. When arrangements for issuing the CFK shares were being made, Jadot told Williams that some modification of the Pooling Agreement would be necessary although he still expected most copper traffic to use the Benguela Railway.[75] In March 1929 the BCK officially informed the Benguela Railway Company that as far as they were concerned the only parts of the agreement which were still in force were those pertaining to the construction of the link between the CFK and the frontier.[76]

Jadot continually assured Williams that the Benguela Railway would get all the traffic it could carry. On the Benguela Railway side there were some very strong doubts on this point, but Williams thought the need to recover the investment made in constructing the link would serve as sufficient guarantee that the Belgians would provide a reasonable amount of traffic; he did not think he could justifiably oppose the cancellation of the pool.[77] In replying officially, Alvaro Machado, who had succeeded his father as Benguela Railway Company Managing Director, said he did not think the effects of the pool could be assessed until it had been in operation for a few years. He noted the BCK's desire to cancel the pool, however, and expressed the hope that the two railway companies would be able to come to some agreement with regard to traffic over their lines so as to maintain friendly relations between them.[78]

Actual construction of the line between Dilolo and Tenke began from the Dilolo end. The Benguela Railway Company, in co-operation with the Angola government, provided the necessary labour for the work[79] and profited considerably from the transport of construction materials. In 1929 working receipts increased over those of 1928 by more than 14,000,000 escudos, almost entirely because of construction traffic. In 1930 the BCK began construction from Tenke, transporting materials over its own line. The Benguela Railway's net earnings—over 21,000,000 escudos in 1929—dropped to less than 5,000 escudos in 1930. As the link line neared completion

[75] 'Extract from Sir Robert Williams' Diary for 14 January 1929', Williams Papers.

[76] BCK to BRC, 20 Mar. 1929, Williams Papers.

[77] Williams to A. Machado, 25 Mar. 1929, Williams Papers.

[78] BRC to BCK, 15 Apr. 1929, Williams Papers.

[79] BR 6, *passim.*

the company's directors continued to express optimism as to the company's future once the line began carrying its full share of Union Minière's production,[80] but decreasing revenues were in fact only a portent of a more serious situation to come.

Williams's dream of joining the Katanga mining area to the west coast of Africa by a continuous line of rail was fulfilled on 10 March 1931 when, at a point 90 miles from Tenke, the last tracks were laid. Until then the actual working of completed sections had virtually no effect on the financial position of either Tanks or the railway company itself. Norton-Griffiths had operated the part of the line he built on his own account until the end of June 1908. Traffic and revenue figures for that period are not available. Figures for the operation of the line from 1908 by the railway company are given in Table 1. Throughout the period 1908–31 carriage of construction materials accounted for a substantial proportion of working receipts, but the marked increase in revenues following the war also reflected increased exports of agricultural products from the Angolan plateau, the fertile region that benefited greatly from the building of the railway, although it was unable to provide sufficient traffic to cover the costs of operating and financing the railway. After the First World War the Zambesia Exploring Company financed several attempts to develop ranching and farming in the area of the land concession, but these met with little success. Extensive coal prospecting was also to no avail; no coal of sufficiently high quality has yet been found in Angola, and fuel for the locomotives (and for the stoves in dining carriages) is provided from forests of fast-growing eucalyptus trees maintained by the railway company. Without large amounts of mineral traffic the railway was not, and had never been expected to be, a paying proposition.

By the time the line was connected to the Katanga mines, the possibility of providing the Benguela Railway with sufficient traffic to get it out of its financial difficulties were remote indeed. After the crash of the New York Stock Market in October 1929, the market for copper deteriorated seriously as the rate of new industrial investment dropped. Copper Exporters Incorporated managed to keep world prices fairly high until April 1930 when increased pressure from consumers finally forced them to allow the price to be cut by 30 per cent. The cartel limited Union Minière's output in 1931 to 120,000 tons, which was cut in 1932 first to 80,000 tons, then to 40,000. Even

[80] BRC Annual Reports for 1929 and 1930.

though prices continued to fall, annual production and stockpiles were not consumed. Even when Union Minière withdrew from Copper Exporters following the imposition by the United States Government of a 4 per cent import duty on copper, output for 1933 was only allowed to reach 66,596 tons.[81] The Benguela Railway was guaranteed a minimum of 20 per cent of Union Minière's output, which, if production had continued to expand as had been anticipated, would probably have been enough for the company to improve its financial position. With production kept so low, the Benguela route rarely got more than its minimum, and its financial difficulties only increased.

By May 1933 the Benguela Railway Company had a total debenture debt of £9,966,960. Tanks held £4,504,000 of this, the general public £3,359,660, and the African Railway Finance Company the remaining £1,802,500. As guarantor for the redemption of and interest on all these debentures, Tanks had paid out a total of £2,545,447 while receiving none of the interest due to her. The financial strain was increased by the fact that at the time Union Minière was paying no dividend. The 30-franc dividend declared in 1932 only applied to privileged shares, none of which Tanks held by then. (Payment of dividends on capital and dividend shares was only resumed in 1936.) To relieve Tanks of the heavy liability of the debenture guarantee, the existing debentures were cancelled; every £20 could be converted to Tanks 'B' Preference shares which had been created specifically for the purpose, worth £15. The £2,515,447 owed Tanks by the railway company in respect of debenture redemption and interest was written off, and a new series of 4 per cent debentures to the value of £9,666,960 issued to Tanks, secured on the railway company's income. The danger of a take-over of the line by debenture holders was averted, and Tanks's financial position very much improved.

In 1938 there was a marked increase in the amount of Katanga traffic carried by the Benguela Railway. Traffic continued to grow gradually until 1952 when another marked rise reflected the carriage of manganese, which became the major Katanga mineral exported through Lobito. It was only in 1956 that Tanks's balance sheet could show a dividend received from the Benguela Railway Company—£273,697 net. Table 4 shows the dividends received since then. The

[81] UMHK Annual reports for 1932 and 1933. See also Elliott, *International Control . . .*, pp. 459–86.

total of £14,487,000 for the 67 years following the formation of the company against an investment of just under £13,000,000 (not including unpaid debenture interest) is hardly evidence of financial success. None the less the railway has served as an important route for copper traffic from Katanga, and to a lesser extent the Copperbelt. The economies of Angola, Katanga, and other countries in central Africa have benefited from the existence of the railway, although Robert Williams and his companies would have been financially better off had it never been built.

CONCLUSION

Katanga was the focal point of myriad converging and conflicting interests, political, economic, and financial, originating in Belgium, Germany, Portugal, Britain, and South Africa. No absolutely clear distinction between political, economic, and financial aims can be drawn, but they were not identical. While governments generally supported attempts by their nationals to profit from the development and exploitation of Africa's resources, support was not given if doing so would have appeared to be in conflict with wider national interests. In cases of conflict, it was political interests that tended to dominate.

At the centre were the Belgians whose primary concerns, both while the Congo was the personal property of their Sovereign and after they had, however reluctantly, taken it over, were to retain control over the country in their own hands, keeping foreign influence —both political and economic—to a minimum, and secure for Belgium as much as possible of the profits to be gained from exploiting the country's wealth. The Belgian government and the companies operating in Katanga may on occasion have disagreed, but no divergence between government and company policy was allowed to become apparent. Williams was able to secure a foothold in Katanga only because Belgian financiers were not initially prepared to invest there, and the extent to which he might be able to control the exploitation of the copperfields was carefully limited. The fact that his share in Union Minière and related companies diminished over the years was primarily because of his inability to continue providing large amounts of capital, not particularly because of the machinations of his 'Belgian banker colleagues'. The Belgians were fundamentally in a position of great strength, as it was they who had ultimate control over the copper deposits and their working. It was

Williams's failure to appreciate this fact fully that was at the root of much of his difficulty.

Williams's position was also weakened by the fact that his interests could not, with only minor exceptions, be identified with any broader British political, economic, or strategic aims, and were in direct conflict with the interests of the British South Africa Company, Southern Rhodesia, and South Africa, all of whom had much greater influence with the British government than he had. Chartered remained in a position of strength, even after the Benguela Railway was completed, because of Union Minière's continuing demand for Wankie coal and coke, which could not be supplied from the Congo or Angola, or, as cheaply, from Europe or America. Although the Foreign Office, either because of the generally wider view it had of British interests, or because of Williams's personal contacts within it, tended to take a more co-operative line regarding the Benguela Railway than other government departments did, the Colonial Office—Imperial—view tended to prevail. It was only when opposition to the railway from the Empire was withdrawn that the desire to alleviate unemployment in Britain was allowed to come to the fore, enabling Williams to secure indirect financial support from the British government.

Like the Belgians, the Portuguese, both before and after the overthrow of the monarchy in 1910, wanted to maintain control over their African colonies, and keep foreign influence to a minimum. Unlike the Belgians, however, they were forced to rely to a much greater extent on outside sources of finance for colonial development. Portugal's interest in these colonies stemmed not only from the desire for economic gain but also from the tangible links they provided with Portugal's past glories as a sea-faring nation, the centre of a far-flung empire, and a country with influence and power in the world. This was in the past, and although it was the Portuguese who stood to gain most directly from the Benguela Railway, they were in no position to exert diplomatic pressure on its behalf, or to support it financially. In any case, the virtually continuous changes of government in Portugal meant that it was impossible to determine and carry through any long-term economic policies for Angola that would have been of much benefit to the railway.

For Germany, confidently expecting to take over most of Angola eventually, the Benguela Railway had much the same potential value as it had for Portugal. Even without the anticipated transfer of territory, a large share in the railway would have given Germany several advantages. She would have had control over most of a transcontinental rail route and would have been in a position, because of her diplomatic and financial strength, to use and expand it to the advantage of her own colonies. She would also have been able to exert pressure on the Portuguese to grant concessions in southern Angola, and on the Belgians to share with her some of the profits from Katanga. This fell far short of the ideal *Mittelafrika*, but was highly attractive none the less. There was, as far as the Germans were concerned, no reason whatever why Williams had to remain involved in the finance or control of the railway at all.

To a certain extent Williams was a victim of circumstances beyond his control. He could certainly have done nothing about the Portuguese political situation, Germany's intentions regarding the Portuguese colonies, or the outbreak of war. None of these would have posed such serious problems as they did if Williams had been able to appreciate the need for the closer identity of his interests with those of the Belgians, and if he had been able to secure immediately the money necessary for the rapid completion of the line. Once Leyland withdrew his support, delaying the start of construction, the way was open for the various forces opposed to the railway to gather force. As construction was further delayed, Williams appeared increasingly weak, with regard to both his ability to secure political and diplomatic support and his command of financial resources. His business abilities proved no match for those of men like Francqui and Jadot, and in his talks with Ryan he appears almost pathetic.

One of Williams's friends and admirers said that 'on him more than on any other man has fallen the mantle of Rhodes as the developer of British Central Africa'.[82] The comparison is not entirely inapt. On the one hand, Williams was never able to exert the power and influence Rhodes did, and was never able to profit greatly from the exploitation of mineral wealth in British Central Africa. Unlike Rhodes he did not have the great wealth derived from Kimberley diamonds and Transvaal gold behind him, and once he and Leyland had parted company, he did not have the support of the business acumen of an Alfred Beit to help him through difficult times. On the other hand, Williams was, like Rhodes, very much a pioneer, a visionary with a tenacity of spirit that minimized obstacles and

[82] Lord Howard of Penrith, *Theatre of Life*, 2 vols. (Hodder & Stoughton, London, 1935), i. 36.

strove to achieve great goals whatever the difficulties. The difference between the two men may be summed up in the fact that two of Britain's colonies for decades bore Rhodes's name, while only a small town in Angola bears Williams's. (The name of the town of Caala was changed to Robert Williams in 1929 in connection with the official opening of the Benguela Railway.) Whatever Williams's weaknesses or the problems he faced, it is clear that he did much for the economic development of Katanga and Angola. Without his vision and tenacity the development of the Katanga copper field would have been at least very much delayed, while the Benguela Railway would almost undoubtedly never have been built. The economy of all of central Africa would have been very different.

TABLE 1

Benguela Railway, passenger, goods, and mineral traffic, 1908–1934

	Passengers		*Goods*		*Minerals*[2]		
Year	*Number*	*Receipts (000 escudos)*	*Tonnage*[1]	*Receipts (000 escudos)*	*Tonnage*	*Receipts (000 escudos)*	*Kilometres run*
1908	25,957	17·9	7,151	n.a.	—	—	n.a.
1909	30,771	20·2	10,107	n.a.	—	—	n.a.
1910	38,449	33·6	18,456	n.a.	—	—	147,535
1911	48,606	49·5	32,863	n.a.	—	—	154,097
1912	50,818	52·7	38,474	n.a.	—	—	207,596
1913	63,333	53·2	47,533	283·6	—	—	226,087
1914	109,308	63·7	35,582	277·4	—	—	242,218
1915	116,224	100·3	42,071	390·7	—	—	267,977
1916	107,104	87·4	43,289	455·0	—	—	303,972
1917	126,612	93·2	71,543	511·7	—	—	350,492
1918	169,864	146·7	104,891	617·9	—	—	389,516
1919	200,254	160·1	117,394	772·2	—	—	390,920
1920	170,303	341·8	107,304	1,655·8	—	—	430,457
1921	159,045	574·7	124,275	2,872·4	—	—	400,322
1922	179,126	922·8	158,800	n.a.	—	—	481,418
1923	213,283	1,815·9	178,800	10,337·3	—	—	472,085
1924	225,458	3,106·7	188,539	18,236·4	—	—	530,058
1925	233,875	2,555·9	189·464	18,109·5	—	—	556,211
1926	227,699	2,293·6	178,641	21,062·3	—	—	596,920
1927	236,735	2,182·3	236,735	24,606·3	—	—	642,777
1928	224,167	2,191·7	224,167	26,447·2	—	—	706,494
1929	240,871	2,666·7	275,331	40,482·2	—	—	964,322
1930	257,984	2,567·6	267,864	24,436·3	—	—	970,060
1931	237,599	3,234·8	247,715	22,202·7	—	—	982,153
1932	183,558	3,746·6	221,044	21,369·8	19,285	7,262·2	925,899
1933	156,272	2,666·1	298,331	21,976·7	29,317	7,173·3	965,502
1934	157,802	2,227·5	271,434	20,827·9	24,460	6,200·2	958,907

Source: Benguela Railway Co. and Tanganyika Concessions Ltd. Annual Reports, various years.

[1] Including mineral traffic, all years.

[2] Not regularly classiffed separately prior to 1932.

TABLE 2

The Proportion of Katanga Copper Passing through Different Ports, 1927–1960

Year	*Matadi*				*Beira*		*Lobito*		*Dar-es-Salaam*		*Total*
	Via Port Francqui		*Via Stanley-ville*								
	1,000 metric tons	%	*1,000 metric tons*	%	*1,000 metric tons*	%	*1,000 metric tons*	%	*1,000 metric tons*	%	*1,000 metric tons*
1927	—	—	—	—	67	73·9	—	—	24	26·1	91
1928	11	9·6	—	—	75	64·9	—	—	29	25·5	115
1929	30	23·7	—	—	76	59·0	—	—	22	17·3	128
1930	42	27·8	—	—	81	53·4	—	—	29	18·8	152
1931	42	38·1	—	—	36	32·3	20	18·1	13	11·5	111
1932	19	36·4	1	2·7	17	34·1	14	26·8	—	—	51
1933	33	41·9	3	3·5	23	28·7	20	25·9	—	—	79
1934	54	49·1	5	4·5	33	30·0	18	16·4	—	—	110
1935	62	51·2	5	4·1	36	29·8	18	14·9	—	—	121
1936	46	46·3	5	4·9	30	30·3	19	18·5	—	—	100
1937	64	42·4	3	2·1	51	33·9	33	21·6	—	—	151
1938	59	44·9	4	2·8	39	29·8	30	22·5	—	—	132
1939	53	43·1	5	4·1	38	30·8	27	22·0	—	—	123
1940	68	43·2	4	2·2	51	32·6	35	22·0	—	—	158
1941	70	40·0	4	2·6	77	44·1	23	13·3	—	—	174
1942	78	45·5	5	2·7	61	35·5	28	16·3	—	—	172
1943	74	45·4	4	2·2	57	35·1	28	17·3	—	—	163
1944	72	43·2	2	1·4	56	33·6	36	21·8	—	—	166
1945	76	43·4	4	2·5	58	33·5	36	20·6	—	—	174
1946	72	49·3	2	1·4	43	29·4	29	19·9	—	—	146
1947	75	48·4	5	3·2	39	25·2	36	23·2	—	—	155
1948	74	46·5	5	3·0	40	25·0	41	25·5	—	—	160
1949	66	45·0	2	1·7	41	27·8	37	25·5	—	—	146
1950	82	47·4	3	1·7	48	27·4	41	23·5	—	—	174
1951	85	45·2	5	2·6	60	31·5	39	20·7	—	—	189
1952	100	48·9	4	2·0	52	25·4	48	23·7	—	—	204
1953	97	45·9	4	1·6	67	31·7	44	20·8	—	—	212
1954	102	44·9	6	2·6	70	30·8	50	21·7	—	—	228
1955	108	45·6	5	2·4	71	29·9	52	22·1	—	—	236
1956	117	46·1	5	1·9	76	30·1	56	21·9	—	—	254
1957	110	46·0	5	2·0	72	30·0	53	22·0	—	—	240
1958	80	39·6	4	1·9	69	34·2	49	24·2	—	—	202
1959	100	40·7	4	1·6	82	33·3	60	24·4	—	—	246
1960	58	20·0	—	—	117	40·0	114	40·0	—	—	289

Source: Huybrechts, '*Les Voies d'accès et d'évacuation du Congo*, p. 63.

TABLE 3

Union Minière Copper Output, 1911–1960 (in metric tons)

Year	*Tonnage*	*Year*	*Tonnage*
1911	998	1936	95,667
1912	2,492	1937	150,467
1913	7,407	1938	123,943
1914	10,722	1939	122,649
1915	14,042	1940	148,829
1916	22,167	1941	162,167
1917	27,462	1942	165,940
1918	20,238	1943	156,850
1919	23,019	1944	163,610
1920	18,962	1945	160,211
1921	30,464	1946	143,855
1922	43,362	1947	150,840
1923	57,886	1948	155,620
1924	85,570	1949	141,399
1925	90,104	1950	175,920
1926	80,639·3	1951	191,959
1927	89,155	1952	205,749
1928	112,456	1953	214,116
1929	136,992	1954	223,791
1930	138,949	1955	234,673
1931	120,000	1956	247,452
1932	54,000	1957	240,280
1933	66,596	1958	235,586
1934	110,085	1959	280,403
1935	107,682	1960	300,675

Source: 'Tanks' and Union Minière Annual Reports.

TABLE 4

Dividends paid to Tanganyika Concessions Ltd. by the Benguela Railway Co., 1954–1970[1]

Year	*Amount*	*Year*	*Amount*
1954	£273,697	1964	£737,888
1955	273,697	1965	922,360
1956	410,546	1966	922,360
1957	821,093	1967	1,076,087
1958	690,686	1968	1,721,739
1959	690,400	1969	739,810
1960	922,360	1970	941,577
1961	922,360	1971	944,584
1962	737,888	Total	14,487,020
1963	737,888		

Source: Tanganyika Concessions Ltd. Annual Reports.

[1] Financial year ending 31 Dec. As Tanks's financial year ends on 31 July, these sums are included in their accounts for the following year.

Select Bibliography

PRIMARY SOURCES

Official Government Records, Great Britain

1. Cabinet Papers:
   a. CAB 23—Cabinet Conclusions.
   b. CAB 24—Cabinet Minutes.
2. Colonial Office:
   a. CO 96—Gold Coast, Original Correspondence.
   b. CO 417—Africa, South; Original Correspondence respecting the British South Africa Company, the South African Republic, Basutoland, Rhodesia, and Portuguese East Africa.
   c. CO 532—Dominions, Original Correspondence.
   d. CO 551—Africa, South; Union of South Africa, Original Correspondence.
3. Foreign Office:
   These were of particular interest and value because Williams called on the Foreign Office for assistance, not only in the same way that many British businessmen operating in foreign countries or colonies did, but also on the basis of his personal relationship with the Grey family. After Sir Edward Grey left the Foreign Office, Williams maintained close contact with a number of permanent officials and was in the habit of sending them copies of letters concerning his affairs as well as lengthy memoranda, many of which have not been preserved in his papers or the files of his companies. A large amount of information is also provided by official and unofficial comments made by various British embassy, legation, and consular officers in Brussels, Lisbon, Angola, and Katanga. The following classes of documents have been consulted:
   a. FO 2—Africa.
   b. FO 84—Slave-Trade.
   c. FO 123—Embassy and Consular Archives, Belgium. For the period 1914–45 no files of this class are known to exist, and were apparently destroyed.
   d. FO 179—Embassy and Consular Archives, Portugal.
   e. FO 367—Africa, New Series.
   f. FO 371—Political (also referred to as Diplomatic).
   g. FO 800—Private Collections.
   This class consists of letters written privately by members of the Foreign Office and Diplomatic Corps. Letters to and from Sir Francis Bertie, Sir Edward Grey, and Lord Lansdowne were used here.

4. India Office:
   a. File No. 779, Judicial and Public Department.
5. Treasury:
   a. T 160—Finance, file No. 4273.

Official Government Records, Belgium

1. The Archives of the État Indépendant du Congo in the African Section of the Foreign Ministry, formerly the Ministry of Colonies:
   a. File RF 1713—Correspondence and notes regarding various negotiations between Williams and the Congo State.
   b. File IRCB 505–7—Correspondence between Baron E. van Eetvelde and King Leopold II.
2. Archives of the Belgian Ministry of Foreign Affairs:
   a. Political files:
      Af 1–1—Africa, General.
      Af 1–9—Africa, Portugal.
      Af 9—Portugal.
3. The papers of King Leopold II in the Palace at Brussels. Those of Leopold's papers that survive dealing with foreign affairs of the Congo have been classified according to the writer, then listed in chronological order. The numbers given these letters in footnotes are those assigned them in the detailed list and index prepared by the royal archivist. Some letters not included in this list have also been cited.

Private Papers, Britain

1. Archives and records held by Tanganyika Holdings Limited:
   These were the most important sources of material for this work, and included:
   a. A partial collection of the personal and business papers of Sir Robert Williams, consisting of:
      1. Sir Robert's office diaries, 1913–38.
      2. A large number of letters, originals and copies, covering virtually all of Williams's activities in Africa. With a few exceptions these have never been properly catalogued and appear to have been collected for the preparation of Williams's memoirs, on which he began work in 1914. On the advice of Sir John Murray, the publisher, Williams secured the ghost-writing assistance of Frederic Swann, a correspondent for *The Times*. Swann produced a partial draft but died in 1923 before completing it. The work was taken over by one F. Dawson who produced a draft of several more chapters before giving up the job. After a lapse of several years the work was undertaken by a Col. McKinnon who wrote a complete draft that Williams was unable to approve before his death in 1938. Godfrey Hutchinson also died before finishing more than a partial version of the memoirs, as did his son Robert. Tanganyika

Holdings then asked Commander George Martelli to complete Robert Hutchinson's manuscript, a task which was finally completed in 1971.

3. Extracts of letters from Williams to his wife.

b. Files of Tanganyika Concessions Limited:

1. TC 24 and 25—This file covers the period 1895–1909 and consists of letters collected and copied for use in the legal proceedings between Christopher Leyland and Tanks. Many of the letters are duplicated elsewhere while many can only be found here.
2. TC 134—Correspondence between Williams and A. A. Thompson, while the latter was in Cape Town.
3. TC 138A—Correspondence with H. G. Watson.
4. TC 139—Correspondence with H. F. Watson.
5. TC 151—Negotiations with Cattier re Preference shares.
6. TFN—Correspondence with the Trade Facilities Act Advisory Committee.

c. Files of correspondence with Union Minière, consisting largely of copies of letters from Katanga on technical subjects:

1. UM 64—Africa Correspondence, General.
2. UM 75/346—Native Labour.
3. UM 75/347—Nyassa Native Labour.
4. UM 76—African Correspondence, General.

d. Files of the Zambesia Exploring Company:

1. ZE 3—Sir Robert Williams & Co., 1901–25.
2. ZE 4—Sir Robert Williams & Co., 1893–1900.
3. ZE 7—Société Générale de Belgique loan.

e. Files of the London Committee of the Benguela Railway Company:

1. BR 6—Works and Construction, 1929–31.
2. BR 21—Finance by Société Générale de Belgique.
3. BR 31—Debenture issue, 1922.

2. The George Cawston Papers, Rhodes House, Oxford.
3. The diaries of Sheffield Neave, particularly those kept during his journey to and in Katanga in 1907–8 and his trip with Jadot in 1911–12.
4. The Philip Lyttelton Gell Papers. At the time these were consulted they were temporarily on deposit at the National Register of Archives, and had not been catalogued. This task has subsequently been completed by Mr. T. Jayne of the National Register's staff. These papers provide some very interesting insights into Chartered's operations and are particularly important as none of the company's London archives are now available for the period covered by this work.
5. The Cecil John Rhodes Papers, Rhodes House, Oxford, Volume 3A, Charters.
6. The papers of General Sir Francis Reginald Wingate.

These are part of the Sudan Archive of Durham University Library's Oriental Section. They are not kept as a separate collection.

7. A summary of negotiations with Schröder & Co. for a loan for the construction of the Benguela Railway. As these negotiations were not fruitful the original correspondence was not kept.

Private Papers, Belgium

1. The papers of Baron Edmund van Eetvelde, Archives Générales du Royaume.
2. Two files made available to me by courtesy of M. Ph. Chevalier, one containing letters to and from Albert Thrys and Gaston Perier, apparently originating in the offices of the CCCI. The second probably came from Jadot's offices in the Société Générale de Belgique.
3. Some private information was given me regarding the Comité Spécial du Katanga.
4. The papers of Albert Thys, Musée Royale de l'Afrique Centrale, Letters to Thys from King Leopold II.

Some of the collections of private papers that I consulted but found of little value were the Morel Papers at the London School of Economics and Political Science, the papers of the 4th Earl Grey at the University of Durham, Department of Palaeography and Diplomatic, and at the Musée Royale de l'Afrique Centrale, the papers of Jules Cornet, and those of Henri Buttgenbach.

SECONDARY SOURCES

General Reference Works:

Biographie coloniale belge.
Colonial (and Dominions) Office list.
Directory of directors.
Encyclopedie coloniale et martime.
Foreign Office list.
Grande enciclopédia Portuguesa e Brasileira.
Register of defunct and other companies removed from the stock exchange official yearbook.
Who was who.
Who's who.

British Government Command Papers:

C 4361, Africa No. 4 (1885), 'Protocols and general act of the West African conference' (Parliamentary Papers, 1884/5, LV, 133). C 4739, Africa No. 3 (1886), 'General act of the conference of Berlin signed February 26 1885' (Parliamentary Papers, 1886, xlvii, 97).

Unpublished Works:

COOKEY, S. J. S., 'Great Britain and the Congo question 1892–1913', University of London, Ph.D. thesis, 1964.

GUBEL, R. 'Proefschrift de Benguella Railway', Mémoire, Coloniale Hogeschool, 1948–9.

MARKS, SALLY J., 'British policy towards Belgium from the armistice to the fall of Lloyd George', University of London, Ph.D. thesis, 1968.

NORTON-GRIFFITHS, GWLADYS, LADY, 'The story of my family and the story of my life', typescript for private circulation.

POGGE VON STRANDMAN, H. J. O., 'Germany, Britain and Africa, 1906–14', Oxford University, D.Phil. thesis, 1970.

Books and Articles:

ANSTEY, ROGER F., *Britain and the Congo in the nineteenth century* (Clarendon Press, Oxford, 1962).

ASCHERSON, NEAL, *The king incorporated* (Allen & Unwin, London, 1963).

BANCROFT, J. AUSTEN, *Mining in Northern Rhodesia*, arranged and prepared by T. D. Guernsey (British South Africa Co., [London] 1961).

BLUNDEN, MARGARET, *The countess of Warwick* (Cassell, London, 1967).

BRAGANÇA-CUNHA, V. DE, *Revolutionary Portugal (1910–1936)* (James Charles, London, [1938]).

Brief History of the Benguela Railway, describing its construction through Angola, Portuguese West Africa, and the important role it is destined to play in the development of southern and central Africa (Benguela Railway Co., London, [1929]).

BUTTGENBACH, H., 'Au Katanga: less mines de Kambove', *Mouvement Géographique*, xix (1902), 577.

CAMERON, V. L., *Across Africa*, 2 vols. (Baldy Ibister & Co., London, 1877).

CAPELLO, H. C., and R. IVENS, *From Benguella to the territory of Yacca; description of a journey into central and west Africa*, translated by Alfred Elwes, 2 vols. (Sampson Lowe, London, 1882).

COLLINS, ROBERT O., *King Leopold, England and the Upper Nile, 1899–1904* (Yale University Press, New Haven, Conn., and London, 1968).

Comité Spécial du Katanga, 1900–1925 (Office de Publicité, Brussels, 1927).

Comité Spécial du Katanga, 1900–1950 (Cuypers, Brussels [1950]).

Compagnie du chemin de fer du Bas-Congo au Katanga, 1906–1956 ([BCK] Brussels [1956]).

Congo Illustré, Brussels, Vols. i-iv (1891–5).

COOKEY, S. J. S., *Britain and the Congo question, 1885–1913* (Longmans, London, 1968) (The published version of this work does not include all the relevant material contained in the thesis).

COPPER DEVELOPMENT ASSOCIATION, *Copper, its ores, mining and extraction* (C.D.A. Publication No. 46, revised ed., 1946).

—— *Introduction to copper* (C.D.A. Publication No. 52, no date).

CORNET, JULES, 'Les Gisements métallifères du Katanga', *Mémoires de la Société des Sciences, Lettres et Art de Hainaut*, viii (1894), 3;

subsequently published in *Revue Universelle des Mines*, 3rd Ser. xxvii (1894), 217, and in *Bulletin de la Société Belge de Géologie, de Paléontologie et d'Hydrologie*, xvii (1903), 3.

CORNET, RENE-JULES, *Katanga: le Katanga avant les Belges et l'expedition Bia-Francqui-Cornet*, 3rd ed. (Cuypers, Brussels, 1946).

CROWE, S. E., *The Berlin West African conference, 1884–1885* (Longmans, Green & Co. for the Royal Empire Society, London, 1942).

D'ALMADA, J. *Para a historia do caminho de ferro de Benguela*, published in English as *An historical outline of the Benguela railway* (Benguela Railway Company, Lisbon, 1951).

DAVIDSON, BASIL, *In the eye of the storm; Angola's people* (Longman, London, 1972).

DAVIS, WATSON, *The story of copper* (Century, New York, 1924).

DEMANGEON, A., 'The Cape-to-Cairo dream', *The Living Age*, cccxxv (1929), 364.

d'ERLANGER, ÉMILE B., *The history of the construction of the Rhodesian transport system* (Burrup Mathisson, London, 1938).

DRECHSLER, HORST, 'L'Allemagne et l'Angola du sud, 1898–1903', *Présence Africaine*, xlii (1962), 54. A slightly different version of this article was published as 'Germany and S. Angola, 1898–1903', in *Actas do Congresso International de Historia des Descobrimentos*, iv (Lisbon, 1961), 73.

DUFFY, JAMES, *Portuguese Africa* (Harvard University Press, Cambridge, Mass., 1959).

EGERTON, F. CLEMENT C., *Angola in perspective; endeavor and achievement in Portuguese West Africa* (Routledge & Kegan Paul, London, 1957).

ELLIOTT, W. Y., *et al.*, *International control in the non-ferrous metals* (Macmillan, New York, 1937).

FRANKEL, S. HERBERT, *Capital investment in Africa, its course and effects* (Oxford University Press, London, 1938).

GANN, L. H., *A History of Northern Rhodesia; early days to 1953* (Chatto & Windus, London, 1964).

—— *A history of Southern Rhodesia, early days to 1934* (Chatto & Windus, London, 1965).

—— 'The Northern Rhodesian copper industry and the world of copper, 1923–1952', *Rhodes-Livingstone Journal*, xviii (1955), 1.

—— *The birth of a plural society; the development of Northern Rhodesia under the British South Africa Company, 1894–1914* (Manchester University Press for the Rhodes-Livingstone Institute, 1958).

GELFAND, MICHAEL, *Northern Rhodesia in the days of the charter; a medical and social study, 1878–1924* (Blackwell, Oxford, 1961).

GIFFORD, PROSSER, and W. R. LOUIS (eds.) with Alison Smith, *Britain and Germany in Africa; imperial rivalry and colonial rule* (Yale University Press, New Haven, Conn., and London, 1967).

GREGORY, SIR THEODORE, *Ernest Oppenheimer and the economic development of Southern Africa* (Oxford University Press, Cape Town, 1962).

GRENVILLE, J. A. S., *Lord Salisbury and foreign policy; the close of the nineteenth century* (Athlone Press, London, 1964).

GUNNING, H. C., 'The early years' in Felix Mendelsohn (ed.), *The geology of the Northern Rhodesia copperbelt* (MacDonald, London, 1961).

HAILEY, VISCOUNT, *An African survey, revised 1956* (Oxford University Press, London, 1957).

HAMMOND, R. J., *Portugal and Africa, 1815–1910; a study in uneconomic imperialism* (Stanford University Press, Stanford, Cal., 1966).

HANCE, W. A., and I. S. VAN DONGEN, 'The port of Lobito and the Benguela Railway', *Geographical Review*, xlvi, No. 4 (1956), 460.

HANCOCK, SIR W. KEITH, *Smuts*, Volume i, *The Sanguine Years, 1870–1919* and Volume ii, *The fields of force* (Cambridge University Press, Cambridge, 1962 and 1968).

HANNA, A. J., *The beginnings of Nyasaland and North-Eastern Rhodesia, 1859–1895* (Clarendon Press, Oxford, 1956).

HATTON, P. H. S., 'Harcourt and Solf: the search for an Anglo-German understanding through Africa, 1912–1914', *European Studies Review*, i, No. 2 (1971), 123.

HENDERSON, W. O., *Studies in German colonial history* (Cass, London, 1962).

HERFINDAHL, ORRIS C., *Copper costs and prices, 1870–1957* (Johns Hopkins University Press, Baltimore, Md., 1959).

HERTSLET, E., *The map of Africa by treaty*, 3 vols., 3rd ed. revised by R. W. Brant (H.M.S.O., London, 1909).

HILL, M. F., *Permanent way; the story of the Kenya and Uganda railways*, 2 vols. (East African Railways and Harbours Board, Nairobi, [1952]).

HOWARD, ESME WILLIAM, Lord Howard of Penrith, *Theatre of Life*, 2 vols. (Hodder & Stoughton, London, 1935).

HUYBRECHTS, ANDRÉ, 'Les Voies d'accès et d'évacuation du Congo', *Cahiers Economiques et Sociaux*, vii (1969), 3.

KATZENELLENBOGEN, S. E. 'Sources for the study of practical imperialism', *Business Archives*, xxxii (1970), 16.

—— 'The miner's frontier, transport and general development' in L. H. Gann and Peter Duignan (eds.), *Colonialism in Africa 1870–1960*, iv (Cambridge University Press, Cambridge, forthcoming).

KEITH, A. B. *The Belgian Congo and the Berlin act* (Clarendon Press, Oxford, 1919).

KRISHNAMURTHY, B. S., 'The Thomson treaties and Johnston's certificates of claim', *African Social Research*, viii (1969), 589.

LANGER, W. L., *The diplomacy of imperialism*, 2nd ed. (Knopf, New York, 1950).

LEMAIRE, CHARLES, 'On the Congo: the Belgian scientific expedition to Ka-Tanga', *Scottish Geographical Magazine*, xvii (1901), 526.

LOCKHART, J. G., and C. M. WOODHOUSE, *Cecil Rhodes; the colossus of southern Africa* (Hodder & Stoughton, London, 1963).

MAGNUS, SIR PHILIP MONTEFIORE, *King Edward the Seventh* (John Murray, London, 1964).

MEREDITH, HUBERT A., *The drama of money making; tragedy and comedy of the London stock exchange* (Sampson Lowe & Marston, London, [1931]).

MOLONEY, J. A., *With Captain Stairs to Katanga* (Sampson Lowe & Marston, London, 1893).

Morphologie des groupes financiers, 2nd ed. (Centre du Recherche et d'Information Socio-Politique, Brussels, no date).

OLIVER, ROLAND, *Sir Harry Johnston and the scramble for Africa* (Chatto & Windus, London, 1957).

OLYFF, JOSEPH, *Le Comité Spécial du Katanga* (Puvrez, Brussels, 1932).

PAM, ALBERT, *Adventures and recollections*, (Oxford University Press for Pam, Oxford, 1945).

PATÉ, OPTAT 'Le Comité Spécial du Katanga', *Congo*, ii (1926), 63.

PAULING, GEORGE, *The chronicles of a contractor, being the autobiography of the late George Pauling*, edited by David Buchan (Constable, London, 1926).

PERIER, G. P., *Les Flèches du Congophile* (Éditions l'Afrique et le Monde, Brussels, 1952),

PONSONBY, FREDERICK, *Recollections of three reigns*, 2 vols. (Eyre & Spottiswoode, London, 1951).

PROUT, COL. H. G., 'The Cape-to-Cairo railway', *Munsey's Magazine*, xxi (1899), 113.

RAPHAEL, LOIS A. G., *The Cape-to-Cairo dream; a study in British imperialism* (Columbia University Press, New York, 1936).

ROBINSON, R. R., and J. A. GALLAGHER with ALICE DENNY, *Africa and the Victorians* (Macmillan, London, 1961).

ROSS, EDWARD ALSWORTH, *Report on employment of native labour in Portuguese Africa* ([Abbot Press], New York, 1925).

ROSSLYN, J. F. H. St. C. ERSKINE, 5TH EARL OF, *My gamble with life* (Cassell, London, 1928).

ROTBERG, ROBERT I., *Africa and its explorers* (Cambridge University Press, Cambridge, 1970).

—— *Joseph Thomson and the exploration of Africa* (Chatto & Windus, London, 1971).

—— 'Plymouth Brethren and the occupation of Katanga', *Journal of African History*, v (1964), 285.

SHARP, R. R., *Early days in Katanga* (No publisher, Bulawayo, 1956).

SHARPE, SIR A., 'The backbone of Africa', *Geographical Journal*, lii (1918), 141.

SHARPE, SIR A., 'A journey to Garenganze', *Proceedings of the Royal Geographical Society*, N.S. xiv (1892), 36.

SKELTON, ALEX, 'Copper' in W. Y. Elliott *et al.*, International control in the non-ferrous metals (Macmillan, New York, 1937).

SLADE, RUTH, *King Leopold's Congo; aspects of the development of race relations in the Congo Independent State* (Oxford University Press for the Institute of Race Relations, London, 1962).

SMITH, ALISON, 'The southern section of the interior, 1840–84', in Roland Oliver and Gervase Mathew (eds.), *History of East Africa*, i (Clarendon Press, Oxford, 1963).

SMUTS, J. C., *Jan Christian Smuts* (Cassell, London, 1952).

Some account of George Grey and his work in Africa by friends who knew him there (Chiswick Press, London, 1914).

SOREL, P., 'Les Origines et les débuts de la Compagnie du chemin de fer du Katanga', in *Comptes Rendus du Congres Scientifique, Elisabethville, 1950, Tome VII, Commission historique, juridique et administrative* (CSK Brussels, [1950]).

STENGERS, JEAN, *Belgique et Congo; l'élaboration de la charte coloniale* (La Renaissance du Livre, Collection Notre Passée, Brussels, 1963).

—— *Combien le Congo a-t-il couté á la Belgique* (Academie Royale des Sciences Coloniales, Classe des Sciences Morales et Politiques Mémoires in 8° Nouvelle Série, Tome XI, Fasc 1 (Histoire), Brussels, 1957).

—— 'The Congo Free State and the Belgian Congo' in L. H. Gann and Peter Duignan (eds.), *Colonialism in Africa, 1870–1960, Volume I, The history and politics of colonialism, 1870–1914* (Cambridge University Press, Cambridge, 1969).

—— 'Leopold II et la fixation des frontiers du Congo', *Le Flambeau*, xlvi, no. 3/4 (1963), 153–97.

—— 'L'Imperialisme colonial de la fin du XIXe siècle; mythe ou réalité', *Journal of African History*, iii (1962), 469.

STENMANS, ALAIN, *La Reprise du Congo par la Belgique; essai d'histoire parlementaire et diplomatique* (Louis, Brussels, 1949).

THOMSON, J. B., *Joseph Thomson, African explorer* (Sampson Lowe & Marston, London, 1896).

THOMSON, JOSEPH, 'To Lake Bangweolo and the unexplored regions of British Central Africa', *Geographical Journal*, i (1893), 97.

THOMSON, R. S. *Fondation de l'État Indépendant du Congo; un chapitre de l'histoire du partage de l'Afrique* (Office de Publicité, Brussels, 1933).

TRUPTIL, R. J. *British banks and the London money market* (Jonathan Cape, London, 1936).

Union Minière du Haut-Katanga, 1906–1956, 2nd ed. (Cuypers, Brussels, 1956).

Union Minière du Haut-Katanga, 1906–1956, évolution des techniques et des activités sociales, titre I, géologie et recherches minières (UMHK, Brussels, 1956).

VANDEWOUDE, ÉMILE, *Inventaire des archives relatives au développement extérieur de la Belgique* (Archives Générales du Royaume, Brussels, 1965).

VAN ZUYLEN, Baron Pierre, *L'Échiquier congolais ou le secret du roi* (Dessart, Brussels, 1959).

VARIAN, H. F., 'Geography of the Benguela Railway', *Geographical Journal*, lxxviii (1931), 497.

—— *Some African milestones* (George Ronald, Wheatley and Oxford, 1953).

VERBEKEN, AUGUST, *M'siri, roi du Garenganze* (Cuypers, Brussels, 1956).

—— and M. Walraet, *La Première Traversée du Katanga en 1806; voyage des 'pombeiros' d'Angola au Rios de Sena* (Institut Royal Colonial Belge, Brussels, 1953).

VERDICK, EDGARD, *Les Premiers Jours au Katanga (1890–1903)* (CSK, Brussels, 1952).

WALKER, ERIC A., *A history of southern Africa*, 3rd ed. (Longmans, London, 1957).

WALLIS, J. P. R., *One man's hand, the story of Sir Charles Coghlan* (Longmans, London, 1950).

WALTZ, H. *Das Konzessionwissen in Belgishchen Kongo*, 2 vols. (Gustav Fischer, Jena, 1917).

WARHURST, P. R. *Anglo–Portuguese relations in South-Central Africa, 1890–1900* (Royal Commonwealth Society, London, 1962).

WARWICK, FRANCES, Countess of, *Afterthoughts* (Collins, London, 1931).

—— *Life's ebb and flow* (Hutchinson, London, 1929).

WHEELER, DOUGLAS L., and RENÉ PELISSIER, *Angola* (Pall Mall Press, London, 1971).

WILLEQUET, J. *Le Congo belge et la Weltpolitik* (Université Libre de Bruxelles, Publications de la Faculté de Philosophie et Lettres, Tome 22, Brussels, 1963).

WILLIAMS, ROBERT, 'German penetration in Central Africa' (Address to the London Chamber of Commerce, 26 June 1918).

—— 'The Cape to Cairo railway', Address to the African Society, 21 April 1921.

—— 'The dark continent: African transport today and tomorrow', A broadcast talk by Sir Robert Williams, Bart., at the request of the B.B.C., 28 November 1930.

—— 'The milestones of African civilization', Paper read before the Royal Colonial Institute, 8 May 1917.

—— 'More milestones of African civilization and some problems', Paper read before the Royal Colonial Institute, 8 January 1924.

WINGATE, SIR RONALD, *Wingate of the Sudan* (John Murray, London, 1955).

Index